Praise for *The Knowledge*

"*The Knowledge* is a fascinating look at the basic principles of the most important technologies undergirding modern society. . . . A fun read full of optimism about human ingenuity." —*The Wall Street Journal*

"[Dartnell's] plans may anticipate the destruction of our world, but embedded in them is the hope that there might be a better way to live in the pre-apocalyptic world we inhabit right now." —*The Boston Globe*

"The ultimate do-it-yourself guide to 'rebooting' human civilization. With scientific nous, Dartnell depicts probable environmental scenarios on a stricken Earth and offers putative survivors instruction in the technologies needed to craft a culture from the ground up. Many will thrill to this reminder of our species' prodigious resilience." —*Nature*

"*The Knowledge* is kin to the 'way things work' books of the artist-writer David Macauley, showing how complex the makings of civilization— the engines, the infrastructure, all the things we take for granted—really are. . . . [It] will prove a valuable owner's guide for a difficult but not impossible future." —*Kirkus Reviews*

"Dartnell's vision is a great start in understanding what it took to build our world." —*Booklist*

"This book is an extraordinary achievement. With lucidity and brevity, Dartnell explains the rudiments of a civilization. It is a great read even if civilization does not collapse. If it does, it will be the sacred text of the new world—Dartnell that world's first great prophet." —*The Times* (London)

"*The Knowledge* is premised on an ingenious sleight of hand. Ostensibly a manual on rebuilding our technological life-support system after a global catastrophe, it is actually a glorious compendium of the knowledge we have lost in the living; the origins of the material fabric of our actual, unapocalyptic lives. . . . The most inspiring book I've read in a long time." —*The Independent* (London)

"A marvelously astounding work: In one graceful swoop, Lewis Dartnell takes our multilayered, interconnected modern world, shows how fragile its scaffolding is, and then lays out a how-to guide for starting over from scratch. Imagine *Zombieland* told by Neil deGrasse Tyson and you'll get some sense of what a delight *The Knowledge* is to read."
—Seth Mnookin, *New York Times* bestselling author of
The Panic Virus and associate director of MIT's
graduate program in science writing

"For all those terrified by runaway climate change, supereruptions, planet-killer asteroids, doomsday viruses, nuclear terrorism, and absolute domination by superintelligent machines, Lewis Dartnell has written a long-overdue guide to what you should do after the apocalypse: an illuminating and entertaining vision of how to reboot life, civilization, and everything. Dartnell's vision of the survival of the smartest in a postapocalyptic world offers a remarkable and panoramic view of how civilization actually works."
—Roger Highfield, journalist, author, and Science
Museum (London) executive

"This book is useful if civilization collapses, and entertaining if it doesn't. After the cometary impact it may save your life, and if it doesn't at least you'll know why you perished."
—S. M. Stirling, *New York Times* bestselling author of
The Given Sacrifice

"Dartnell makes the technology and science of everyday life in our civilization fascinating and understandable. This book may or may not save your life but it'll certainly make it more interesting. This is the book we all wish we'd been given at school: *The Knowledge* that makes everything else make sense." —Ken MacLeod, author of *Intrusion* and *Descent*

ABOUT THE AUTHOR

Dr. Lewis Dartnell is a UK Space Agency research fellow at the University of Leicester and writes regularly for *New Scientist*, *BBC Focus*, *BBC Sky at Night*, and *Cosmos*, as well as for newspapers including the (London) *Times*, the *Guardian*, and the *New York Times*. He has won several awards, including the Daily Telegraph Young Science Writer Award. He also makes regular TV appearances and has been featured on *BBC Horizon*, *Stargazing Live*, *Sky at Night*, and numerous times on the Discovery and Science channels. His scientific research is in the field of astrobiology, focusing on how microorganisms might survive on the surface of Mars and the best ways to detect signs of ancient Martian life.

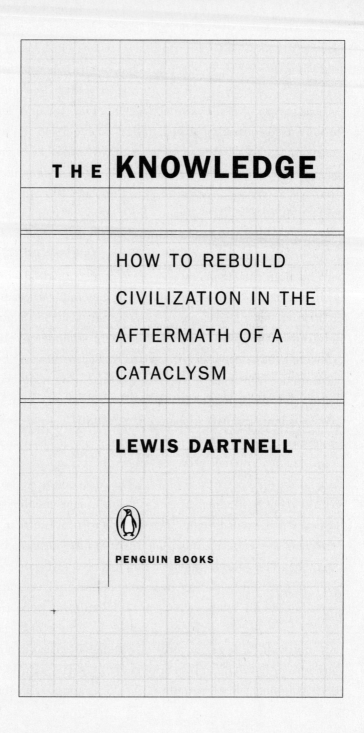

THE **KNOWLEDGE**

HOW TO REBUILD
CIVILIZATION IN THE
AFTERMATH OF A
CATACLYSM

LEWIS DARTNELL

PENGUIN BOOKS

PENGUIN BOOKS

Published by the Penguin Group
Penguin Group (USA) LLC
375 Hudson Street
New York, New York 10014

USA | Canada | UK | Ireland | Australia
New Zealand | India | South Africa | China
penguin.com
A Penguin Random House Company

First published in the United States of America by The Penguin Press,
a member of Penguin Group (USA) LLC, 2014
Published in Penguin Books 2015

THE LIBRARY OF CONGRESS HAS CATALOGED
THE HARDCOVER EDITION AS FOLLOWS:
Dartnell, Lewis.
  The knowledge : how to rebuild our world from scratch / Lewis Dartnell.
    pages cm
  Includes bibliographical references and index.
  ISBN 978-1-59420-523-1 (hc.)
  ISBN 978-0-14-312704-8 (pbk.)
  1. Technology—Popular works. 2. Discoveries in science—Popular works.
  3. Survival—Popular works. 4. Knowledge, Theory of—Popular works.
  I. Title.
  T47.D37 2014
  500—dc23   2013040820

Printed in the United States of America
10 9 8 7 6 5 4 3 2 1

Designed by Marysarah Quinn

The information contained in this book cannot replace sound judgment and good decision making,
which can help reduce risk exposure, nor does the scope of this book allow for disclosure of all the
potential hazards and risks involved. The author and publisher are not responsible for the instructions
and information, as these are not intended for use except in the event of mass disasters, when
the customary ways of doing things are not possible.

To my wife, Vicky.

Thank you for saying yes.

These fragments I have shored against my ruins

T. S. ELIOT, *The Waste Land*

CONTENTS

THE **KNOWLEDGE**

THE WORLD AS WE KNOW IT HAS ENDED.

A particularly virulent strain of avian flu finally breached the species barrier and hopped successfully to human hosts, or was deliberately released in an act of bioterrorism. The contagion spread devastatingly quickly in the modern age of high-density cities and intercontinental air travel, and killed a large proportion of the global population before any effective immunization or even quarantine orders could be implemented.

Or tensions between India and Pakistan reached the breaking point and a border dispute escalated beyond all rational limits, culminating in the use of nuclear weapons. The warheads' distinctive electromagnetic pulses were detected by defense surveillance in China and triggered a round of preemptive launches against the United States, which in turn spurred retaliatory strikes by America and its allies in Europe and Israel. Major cities worldwide were reduced to jagged plains of radioactive glass. The enormous volumes of dust and ash injected into the atmosphere reduced the amount of sunlight reaching the ground, causing a decades-long nuclear winter, the collapse of agriculture, and global famine.

Or the event was entirely beyond human control. A rocky asteroid, only around a mile across, slammed into the Earth and fatally changed atmospheric conditions. People within a few hundred kilometers of ground zero were dispatched in an instant by the blast wave of intense heat and pressure, and from that point on most of the rest of humanity was living on borrowed time. It didn't really matter which nation was struck: the rock and dust hurled up high into the atmosphere—as well as the smoke produced by widespread fires ignited by the heat blast—dispersed on the winds to smother the entire planet. As in a nuclear winter, global temperatures dropped enough to cause worldwide crop failures and massive famine.

This is the stuff of so many novels and films featuring post-apocalyptic worlds. The immediate aftermath is often—as in *Mad Max* or Cormac McCarthy's novel *The Road*—portrayed as barren and violent. Roving bands of scavengers hoard the remaining food and prey ruthlessly on those less well organized or armed. I suspect that, at least for a period after the initial shock of collapse, this might not be too far from the truth. I'm an optimist, though: I think morality and rationality would ultimately prevail, and settlement and rebuilding begin.

The world as we know it has ended. The crucial question is: now what?

Once the survivors have come to terms with their predicament—the collapse of the entire infrastructure that previously supported their lives—what can they do to rise from the ashes to ensure they thrive in the long term? What crucial knowledge would they need to recover as rapidly as possible?

This is a survivors' guidebook. Not one just concerned with keeping people alive in the weeks after the Fall—plenty of handbooks have been written on survival skills—but one that teaches how to orchestrate the rebuilding of a technologically advanced civilization. If you suddenly found yourself without a working example, could you explain how to build an internal combustion engine, or a clock, or a microscope? Or, even more basic, how to successfully cultivate crops and make clothes? The apocalyptic scenarios I'm presenting here are also the starting point for a thought experiment: they are a vehicle for examining the fundamentals of science and technology, which, as knowledge becomes ever more specialized, feel very remote to most of us.

People living in developed nations have become disconnected from the everyday processes of civilization that support them. Individually, we are astoundingly ignorant of even the basics of the production of food, shelter, clothes, medicine, materials, or vital substances. Our survival skills have atrophied to the point that much of humanity would be incapable of sustaining itself if the life-support system of modern civilization failed, if food no longer magically appeared on store shelves, or clothes on hangers. Of course, there was a time when everyone was a survivalist, with a far more intimate connection to the land and methods of production, and to survive in a post-apocalyptic world you'd need to turn back the clock and relearn these core skills.[1]

What's more, each piece of modern technology we take for granted requires an enormous support network of other technologies. There's much more to making an iPhone than knowing the design and materials of each of its components. The device sits as the capstone on the very tip of a vast pyramid of enabling technologies: the mining and refining of the rare element indium for the touch screen, high-precision photolithographic manufacturing of microscopic circuitry in the computing processor chips, and the incredibly miniaturized components in the microphone, not to mention the network of cell phone towers and other infrastructure necessary to maintain telecommunications and the functioning of the phone. The first generation born after the Fall would find the internal mechanisms of a modern phone absolutely inscrutable, the pathways of its microchip circuits invisibly small to the

---

1 Similar, small-scale scenarios have occurred in recent history: with the fall of the Soviet Union in 1991, the small republic of Moldova experienced a crippling crash in its economy, forcing people to become self-sufficient, readopting museum-exhibit technology such as spinning wheels, hand looms, and butter churns.

human eye and their purpose utterly mysterious. The sci-fi author Arthur C. Clarke said in 1961 that any sufficiently advanced technology is indistinguishable from magic. In the aftermath of the Fall, the rub is that this miraculous technology would have belonged not to some starfaring alien species, but to people just a generation in our own past.

Even quotidian artifacts of our civilization that aren't particularly high-tech still require a diversity of raw materials that must be mined or otherwise gathered, processed in specialized plants, and assembled in a manufacturing facility. And all of this in turn relies on electrical power stations and transport over great distances. This point is made very eloquently in Leonard E. Read's 1958 essay written from the perspective of one of our most basic tools, "I, Pencil." The astounding conclusion is that because the sourcing of raw materials and the methods of production are so dispersed, there is not a single person on the face of the Earth who knows how to make even this simplest of implements.

A potent demonstration of the gulf that now separates our individual capabilities and the production of even simple gizmos in our everyday life was offered by Thomas Thwaites when, in 2008, he attempted to make a toaster from scratch while studying for his MA at the Royal College of Art. He reverse-engineered a cheap toaster down to its barest essentials—iron frame, mica-mineral insulating sheets, nickel heating filaments, copper wires and plug, and plastic casing— and then sourced all the raw materials himself, digging them out of the ground in quarries and mines. He also looked up simpler, historical metallurgical techniques, referring to a sixteenth-century text to build a rudimentary iron-smelting furnace using a metal trash can, barbecue coals, and a leaf blower for bellows. The finished model is satisfyingly primitive but also grotesquely beautiful in its own right and neatly underscores the core of our problem.

Of course, even in one of the extreme doomsday scenarios, groups of survivors would not need to become self-sufficient immediately. If the great majority of the population succumbed to an aggressive virus, there would still be vast resources left behind. The supermarkets would remain stocked with plentiful food, and you could pick up a fine new set of designer clothes from the deserted department stores or liberate from the showroom the sports car you've always dreamed about. Find an abandoned mansion, and with a little foraging it wouldn't be too hard to salvage some mobile diesel generators to keep the lighting, heating, and appliances running. Underground lakes of fuel remain beneath gas stations, sufficient to keep your new home and car functioning for a significant period. In fact, small groups of survivors could probably live pretty comfortably in the immediate aftermath of the Fall. For a while, civilization could coast on its own momentum. The survivors would find themselves surrounded by a wealth of resources there for the taking: a bountiful Garden of Eden.

But the Garden is rotting.

Food, clothes, medicines, machinery, and other technology inexorably decompose, decay, deteriorate, and degrade over time. The survivors are provided with nothing more than a grace period. With the collapse of civilization and the sudden arrest of key processes— gathering raw materials, refining and manufacturing, transportation and distribution—the hourglass is inverted and the sand steadily drains away. The remnants provide nothing more than a safety buffer to ease the transition to the moment when harvesting and manufacturing must begin anew.

A REBOOT MANUAL

The most profound problem facing survivors is that human knowledge is collective, distributed across the population. No one individual knows enough to keep the vital processes of society going. Even if a skilled technician from a steel foundry survived, he would only know the details of *his* job, not the subsets of knowledge possessed by other workers at the foundry that are vital for keeping it running—let alone how to mine iron ore or provide electricity to keep the plant operating. The most visible technology we use daily is just the tip of a vast iceberg—not only in the sense that it's based on a great manufacturing and organizational network that supports production, but also because it represents the heritage of a long history of advances and developments. The iceberg extends unseen through both space and time.

So where would survivors turn? A great deal of information will certainly remain in the books gathering dust on the shelves of the now-deserted libraries, bookshops, and homes. The problem with this knowledge, however, is that it isn't presented in a way appropriate for helping a fledgling society—or an individual without specialist training. What do you think you'd understand if you just pulled a medical textbook off the shelf and flipped through the pages of terminology and drug names? University medical textbooks presuppose a huge amount of prior knowledge, and are designed to work alongside teaching and practical demonstrations from established experts. Even if there were doctors among the first generation of survivors, they'd be severely limited in what they could accomplish without test results or the cornucopia of modern drugs they were trained to use—drugs that would be degrading on pharmacy shelves or in defunct hospital storage refrigerators.

Much of this academic literature would itself be lost, perhaps to fires ripping unchecked through empty cities. Even worse, much of the wealth of new knowledge generated each year, including that which I and other scientists produce and consume in our own research, is not recorded on any durable medium at all. The cutting edge of human understanding exists primarily as ephemeral bits of data: as specialist journals' academic "papers" stored on website servers.

And the books aimed at general readers wouldn't be much more help. Can you imagine a group of survivors who had access to only the selection of books stocked in an average store? How far would a civilization get trying to rebuild itself from the wisdom contained in the pages of self-help guides to succeeding in business management, thinking yourself thin, or reading the body language of the opposite sex? The most absurd nightmare would be a post-apocalyptic society discovering a few yellowed and crumbly books and, thinking them the scientific wisdom of the ancients, trying to apply homeopathy to curb a plague or astrology to forecast harvests. Even the books in the science section would offer little help. The latest pop-sci page-turner may be engagingly written, make clever metaphorical use of everyday observations, and leave the reader with a deeper understanding of some new research, but it probably won't yield much pragmatic knowledge. In short, the vast majority of our collective wisdom would not be accessible—at least in a usable form—to the survivors of a cataclysm. So how best to help the survivors? What key information would a guidebook need to deliver, and how might it be structured?

I'm not the first person to wrestle with this question. James Lovelock is a scientist with a formidable track record for striking at the heart of an issue long before his peers. He is most famous for his Gaia hypothesis, which posits that the entire planet—a complex assemblage of rocky crust and oceans and swirling atmosphere, along with the thin smear of life that has established itself across the surface—can be understood as a single entity that acts to damp down instabilities and self-regulate its environment over billions of years. Lovelock is deeply concerned that one element of this system, *Homo sapiens*, now has the capacity to disrupt these natural checks and balances with devastating effect.

Lovelock draws on a biological analogy to explain how we might safeguard our heritage: "Organisms that face desiccation often encapsulate their genes in spores so that the information for their renewal is carried through the drought." The human equivalent envisaged by Lovelock is a book for all seasons, "a primer on science, clearly written and unambiguous in its meaning—a primer for anyone interested in the state of the Earth and how to survive and live well on it." What he proposes is a truly massive undertaking: recording the complete assemblage of human knowledge in a huge textbook—a document that you could, at least in principle, read from cover to cover, and then walk away knowing the essentials of everything that is now known.

In fact, the idea of a "total book" has a much longer history. In the past, encyclopedia compilers appreciated far more acutely than we do today the fragility of even great civilizations, and the exquisite value of the scientific knowledge and practical skills held in the minds of the population that evaporate once the society collapses. Denis Diderot explicitly regarded his *Encyclopédie*, published between 1751 and 1772, as a safe repository of human knowledge, preserving it for posterity in case of a cataclysm that snuffs our civilization as the ancient cultures of the Egyptians, Greeks, and Romans had all been lost, leaving behind only random surviving fragments of their writing. In this way, the encyclopedia becomes a time capsule of accumulated knowledge, all of it arranged logically and cross-referenced, protected against the erosion of time in case of a widespread catastrophe.

Since the Enlightenment our understanding of the world has increased exponentially, and the task of compiling a complete compendium of human knowledge would be orders of magnitude harder today. The creation of such a "total book" would represent a modern-era pyramid-building project, consuming the full-time exertion of tens of thousands of people over many years. The purpose of this toil would be to ensure not the safe passage of a pharaoh to eternal bliss in the afterworld, but the immortality of our civilization itself.

Such an all-consuming undertaking is not inconceivable if the will is there. My parents' generation worked hard to put the first man on the moon: at its peak the Apollo program employed 400,000 people and consumed 4 percent of the total American federal budget. Indeed, you might think that the perfect compendium of current human knowledge has already been created by the phenomenal combined effort of the committed volunteers behind Wikipedia. Clay Shirky, an expert on the sociology and economics of the Internet, has estimated that Wikipedia currently represents around 100 million man-hours of devoted effort in writing and editing. But even if you could print Wikipedia in its entirety, its hyperlinks replaced by cross-referenced page numbers, you'd still be a far cry from a manual enabling a community to rebuild civilization from scratch. It was never intended for anything like this purpose, and lacks practical details and the organization for guiding progression from rudimentary science and technology to more advanced applications. Moreover, a hard copy would be unfeasibly large—and how could you ensure post-apocalyptic survivors would be able to get hold of a copy?

In fact, I believe you can help society recover much better by taking a slightly more elegant approach.

The solution can be found in a remark made by physicist Richard Feynman. In hypothesizing about the potential destruction of all scientific knowledge and what might be done about it, he allowed himself a single statement, to be transmitted securely to whichever intelligent creatures emerged after the cataclysm: What sentence holds the most information in the fewest words? "I believe," said Feynman, "it is the *atomic hypothesis* . . . that *all things are made of atoms—little particles that move around in perpetual motion, attracting each other when they are a little distance apart, but repelling upon being squeezed into one another.*"

The more you consider the implications and testable hypotheses emerging from this simple statement, the more it unfurls to release further revelations about the nature of the world. The attraction of particles explains the surface tension of water, and the mutual repulsion of atoms in close proximity explains why I don't fall straight through the café chair I'm sitting on. The diversity of atoms, and the compounds produced by their combinations, is the key principle of chemistry. This single, carefully crafted sentence encapsulates a huge density of information, which unravels and expands as you investigate it.

But what if your word count wasn't quite so restricted? If allowed the luxury of being more expansive while retaining the guiding principle of providing key, condensed knowledge to accelerate rediscovery, rather than attempting to write a complete encyclopedia of modern understanding, is it feasible to write a single volume that would constitute a survivor's quick-start guide to rebooting technological society?

I think that Feynman's single sentence can be improved upon in a fundamentally important way. Possessing *pure* knowledge alone with no means to exploit it is impotent. To help a fledgling society pull itself up by its own bootstraps, you've also got to suggest how to *utilize* that knowledge, to show its practical applications. For the survivors of a recent apocalypse, the immediate practical applications are essential. Understanding the basic theory of metallurgy is one thing, but using the principles to scavenge and reprocess metals from the dead cities, for instance, is another. The exploitation of knowledge and scientific principles is the essence of technology, and as we'll see in this book, the practices of scientific research and technological development are inextricably intertwined.

Inspired by Feynman, I'd argue that the best way to help survivors of the Fall is not to create a comprehensive record of all knowledge, but to provide a guide to the basics, adapted to their likely circumstances, as well as a blueprint of the techniques necessary to rediscover crucial understanding for themselves—the powerful knowledge-generation machinery that is the scientific method. The key to preserving civilization is to provide a condensed seed that will readily unpack to yield the entire expansive tree of knowledge, rather than attempting to document the colossal tree itself. Which fragments, to paraphrase T. S. Eliot, are best shored against our ruins?

The value of such a book is potentially enormous. What might have happened in our own history if the classical civilizations had left condensed seeds of their accumulated knowledge? One of the major catalysts for the Renaissance in the fifteenth and sixteenth centuries was the trickle of ancient learning back into Western Europe. Much of this knowledge, lost with the fall of the Roman Empire, was preserved and propagated by Arab scholars carefully translating and copying texts; other manuscripts were rediscovered by European scholars. But what if these treatises on philosophy, geometry, and practical mechanisms had been preserved in a distributed network of time capsules? And similarly, with the right book available, could a post-apocalyptic Dark Ages be averted?[2]

ACCELERATED DEVELOPMENT

During a reboot, there's no reason to retrace the original route to scientific and technological sophistication. Our path through history has been long and tortuous, stumbling in a largely haphazard manner, chasing red herrings and overlooking crucial developments for long periods. But with 20/20 hindsight, knowing what we know now, could we give directions straight to crucial advances, taking shortcuts like an experienced navigator? How might we chart an optimal route through the vastly interlinked network of scientific principles and enabling technologies to accelerate progress as much as possible?

Key breakthroughs in our history are often serendipitous—they

---

2 If you ignore the remnant material that will be left behind by the collapse of our society, this thought experiment on aiding the recovery of post-apocalyptic survivors could also provide the manual you'd need to develop a technological civilization from scratch after accidentally falling through a time warp into the Paleolithic era ten thousand years in the past, or crash-landing a spaceship on an uninhabited but clement Earth-like planet. This is the ultimate Robinson Crusoe or Swiss Family Robinson shipwreck fantasy—not washing up on a small deserted island, but starting again on a whole empty world.

were stumbled upon by chance. Alexander Fleming's discovery of the antibiotic properties of *Penicillium* mold in 1928 was a chance occurrence. And indeed, the observation that first hinted at the deep coupling between electricity and magnetism—the twitching of compass needles left next to a wire carrying current—was fortuitous, as was the discovery of X-rays. Many of these key discoveries could just as easily have happened earlier, some of them substantially so. Once new natural phenomena have been discovered, progress is driven by systematic and methodical investigation to understand their workings and quantify their effects, but the initial uncovering can be targeted with a few choice hints to the recovering civilization on where to look and which investigations to prioritize.

Likewise, many inventions seem obvious in retrospect, but sometimes the time of emergence of a key advance or invention doesn't appear to have followed any particular scientific discovery or enabling technology. For the prospects of a rebooting civilization, these cases are encouraging because they mean the quick-start guide need only briefly describe a few central design features for the survivors to figure out exactly how to re-create some key technologies. The wheelbarrow, for instance, could have occurred centuries before it actually did—if only someone had thought of it. This may seem like a trivial example, combining the operating principles of the wheel and the lever, but it represents an enormous labor saver, and it didn't appear in Europe until millennia after the wheel (the first depiction of a wheelbarrow appears in an English manuscript written about 1250 AD).

Other innovations have such wide-ranging effects, aiding a great diversity of other developments, that you would want to beeline directly toward them to support many other elements of the post-apocalyptic recovery. The movable-type printing press is one such gateway technology that accelerated development and had incomparable social ramifications in our history. With a little guidance, mass-produced books could reappear early in the rebuilding of a new civilization, as we'll see later.

And when developing new technologies, some steps in the progression could be skipped altogether. The quick-start guide could aid a recovering society by showing how to leapfrog straight over intermediate stages from our history to more advanced, yet still achievable, systems. There are a number of encouraging cases of this kind of technological leapfrogging in the developing nations in Africa and Asia today. For example, many remote communities unconnected to power grids are receiving solar-power infrastructure, hopping over centuries of the Western progression dependent on fossil fuels. Villagers living in mud huts in many rural parts of Africa are leapfrogging straight to mobile phone communications, bypassing intermediate technologies such as semaphore towers, telegraphs, or land-line telephones.

But perhaps the most impressive feat of leapfrogging in history was achieved by Japan in the nineteenth century. During the Tokugawa shogunate, Japan isolated itself for two centuries from the rest of the world, forbidding its citizens to leave or foreigners to enter, and permitting only minimal trade with a select few nations. Contact was reestablished in the most persuasive manner in 1853 when the US Navy arrived in the Bay of Edo (Tokyo) with powerfully weaponized steam-powered warships, far superior to anything possessed by the technologically stagnant Japanese civilization. The shock of realization of this technological disparity triggered the Meiji Restoration. Japan's previously isolated, technologically backward feudal society was transformed by a series of political, economic, and legal reforms, and foreign experts in science, engineering, and education instructed the nation how to build telegraph and railroad networks, textile mills and factories. Japan industrialized in a matter of decades, and by the time of the Second World War was able to take on the might of the US Navy that had forced this process in the first place.

Could a preserved cache of appropriate knowledge allow a post-apocalyptic society to similarly achieve a rapid developmental trajectory?

Unfortunately, there are limits to how far ahead you can push a civilization by skipping intermediate stages. Even if the post-apocalyptic scientists fully understand the basis underlying an application and have produced a design that would work in principle, it may still be impossible to build a working prototype. I call this the Da Vinci effect. The great Renaissance inventor generated endless designs for mechanisms and contraptions, such as his fantastic flying machines, but few of them were ever realized. The problem was largely that Da Vinci was too far ahead of his time. Correct scientific understanding and ingenious designs aren't sufficient: you also need a matching level of sophistication in construction materials with the necessary properties and available power sources.

So the trick for a quick-start guide must be to provide appropriate technology for the post-apocalyptic world, in the same way that aid agencies today supply suitable intermediate technologies to communities in the developing world. These are solutions that offer a significant improvement on the status quo—an advance from the existing, rudimentary technology—but which are still able to be repaired and maintained by local workmen with the practical skills, tools, and materials available. Thus the aim for an accelerated reboot of civilization is to jump directly to a level that saves centuries of incremental development, but that can still be achieved with rudimentary materials and techniques—the *sweet-spot* intermediate technology.

It is these features of our own history—serendipitous discoveries, inventions that were not waiting for any prerequisite knowledge, gateway technologies that stimulated progress in many areas, and opportunities to leapfrog over intermediate stages—that give us optimism that a well-designed quick-start manual for civilization could give directions toward the most fertile investigations and the crucial principles behind key technologies, guiding an optimal route through the web of science and technology, and so greatly accelerate rebuilding. Imagine science when you're not fumbling around in the dark, but your ancestors have equipped you with a flashlight and a rough map of the landscape.

If a rebooting civilization is not required to follow our own idiosyncratic path of progress, it will experience a completely different sequence of advances. Indeed, rebooting along the same trajectory that our current civilization followed may now be very difficult. The Industrial Revolution was powered largely by fossil energy. Most of these easily accessible fossil energy sources—deposits of coal, oil, and natural gas—have now been mined toward depletion. Without access to such readily available energy, how could a civilization following ours haul itself through a second industrial revolution? The solution, as we'll see, will lie in an early adoption of renewable energy sources and careful recycling of assets—sustainable development will likely be forced on the next civilization out of sheer necessity: a green reboot.

In the process, unfamiliar combinations of technologies will emerge over time. We will take a look at examples of where a recovering society is likely to take a different trajectory in its development—the path not traveled—as well as utilizing technological solutions that for us have fallen by the wayside. To us, Civilization 2.0 might look like a mishmash of technologies from different eras, not unlike the genre of fiction known as steampunk. Steampunk narratives are set in an alternative history that has followed a different pattern of development and is often characterized by a fusion of Victorian technology with other applications. A post-apocalyptic reboot with very different rates of progress in separate fields of science and technology is likely to lead to such an anachronistic patchwork.

CONTENTS

A reboot manual would work best on two levels. First, you need a certain amount of practical knowledge handed to you on a plate, so as to recover a base level of capability and a comfortable lifestyle as quickly as possible, and to halt further degeneration. But you also need to nurture the recovery of scientific investigation and provide the most worthwhile kernels of knowledge to begin exploring.[3]

We'll start with the basics and see how you can provide the fundamental elements of a comfortable life for yourself after the Fall: sufficient food and clean water, clothes and building materials, energy and essential medicines. There will be a number of immediate concerns for the survivors: cultivable crops must be gathered from farmland and seed caches before they die and are lost; diesel can be rendered from biofuel crops to keep engines running until the machinery fails, and parts can be scavenged to reestablish a local power grid. We'll look at how best to cannibalize components and scavenge materials from the detritus of the dead civilization: the post-apocalyptic world will demand ingenuity in repurposing, tinkering, and jury-rigging.

Once the essentials are in place, I'll explain how to reinstate agriculture and safely preserve a stockpile of food, and how plant and animal fibers can be turned into clothes. Materials such as paper, ceramic pottery, brick, glass, and wrought iron are today so commonplace that

---

3 While the most discernible features of a society may be its grand monuments, or art, music, or other cultural output, the basics supporting civilization are fundamentals such as agricultural productivity, sewage treatment, and chemical synthesis. This book will focus on the critical science and technology as they are universal: a particular physical law is true no matter where (or when) you are, and a society even thousands of years in the future will have the same basic needs that can be alleviated by technology—food, clothes, power, transport, and so on. Art, literature, and music are an important part of our cultural heritage, but the recovery of civilization wouldn't be held back half a millennium without them, and the post-apocalyptic survivors will develop their own expressions that hold relevance to them.

they are considered prosaic and boring—but how could you actually make them if you needed to? Trees yield an enormous amount of remarkably useful stuff: from timber material for construction to charcoal for purifying drinking water, as well as providing a fiercely burning solid fuel. A whole range of crucial compounds can be baked out of wood, and even ashes contain a substance (called potash) needed for making essential items such as soap and glass, as well as producing one of the ingredients of gunpowder. With basic know-how you can extract a great deal of other critically useful substances from your natural surroundings—soda, lime, ammonia, acids, and alcohol—and start a post-apocalyptic chemical industry. And as your capabilities recover, the quick-start guide will help the development of explosives suitable for mining and for demolishing the carcasses of ancient buildings, as well as the production of artificial fertilizer, and of the light-sensitive silver compounds used in photography.

In later chapters we'll see how to relearn medicine, harness mechanical power, master the generation and storage of electricity, and assemble a simple radio set. And since *The Knowledge* contains information on how to make paper, ink, and a printing press, the book itself contains the genetic instructions for its own reproduction.

How much can one book invigorate our understanding of the world? I obviously can't begin to pretend this single volume represents a complete documentation of the sum total of human knowledge of science and technology. But I think it provides enough of a grounding in the fundamentals to help survivors in the early years after a Fall, and broad directions for tracing an optimal route through the web of science and technology for a greatly accelerated recovery. And, following the principle of providing condensed kernels of knowledge that unravel under investigation, a single volume can encapsulate a vast treasure trove of information. By the time you put down this manual, you'll understand how to rebuild the infrastructure for a civilized lifestyle. You'll also, I hope, have a firmer grip on some of the beautiful fundamentals of science itself. Science is not a collection of facts and figures: it is the method you need to apply to confidently work out how the world works.

The purpose of a quick-start guide is to ensure that the fire of curiosity, of inquiry and exploration, continues to burn fiercely. The hope is that even in the maw of a cataclysmic shock the thread of civilization is not broken and the surviving community does not regress too far or stagnate; that the core of our society can be preserved; and that these crucial kernels of knowledge, nurtured in the post-apocalyptic world, will flourish once again.

This is the blueprint for a rebooting civilization—but also a primer on the fundamentals of our own.

CHAPTER 1

THE END OF THE WORLD AS WE KNOW IT

The most glorious moment for a work of this sort
would be that which might come immediately in
the wake of some catastrophe so great as to
suspend the progress of science, interrupt the
labors of craftsmen, and plunge a portion of our
hemisphere into darkness once again.

DENIS DIDEROT, *Encyclopédie* (1751–1772)

THE SEEMINGLY OBLIGATORY SCENE in any disaster movie is a panning
shot across a broad highway gridlocked with tightly packed vehicles
attempting to flee the city. Instances of extreme road rage flare as drivers grow increasingly desperate, before abandoning their cars among
the others already littering the shoulders and lanes and joining the
droves of people pushing onward on foot. Even without an immediate
hazard, any event that disrupts distribution networks or the electrical
grid will starve the cities' voracious appetite for a constant influx of
resources and force their inhabitants out in a hungry exodus: mass mi-
grations of urbanite refugees swarming into the surrounding countryside to scavenge for food.

I don't want to get stuck in the philosophical quagmire of debating whether mankind is intrinsically evil or not, and whether a controlling authority is a necessary construct to impose a set of laws and maintain order through the threat of punishment. But it is clear that with the evaporation of centralized governance and a civil police force, those with ill intentions will seize the opportunity to subjugate or exploit those more peaceful or vulnerable. And once the situation seems sufficiently dire, even previously law-abiding citizens will resort to whatever action is necessary to provide for and protect their own families. To ensure your own survival you may have to forage and scavenge for what you need: a polite euphemism for looting.

Part of the glue that binds societies together is the expectation that the pursuit of short-term gains through deception or violence is far outweighed by the long-term consequences. You'll be caught and socially stigmatized as an untrustworthy partner or punished by the state: cheats don't prosper. This tacit agreement between the individuals in a society to cooperate and behave for the collective good, sacrificing a certain amount of their own personal freedom in exchange for benefits such as the mutual protection offered by the state, is known as the social contract. It is the very foundation of all collective endeavor, production, and economic activity of a civilization, but the structure begins to strain and social cohesion loosens once individuals perceive greater personal gains in cheating, or suspect that others will cheat them.

During a severe crisis the social contract can snap altogether, precipitating a complete disintegration of law and order. We need look no further than the most technologically advanced nation on the planet to see the effects of a localized fracture in the social contract. New Orleans was physically devastated by the rampage of Hurricane Katrina, but it was the desperate realization by the city's inhabitants that local governance had evaporated and no help would be arriving anytime soon that precipitated the rapid degeneration of normal social order and the outbreak of anarchy.

So after a cataclysmic event, we might expect organized gangs to emerge to fill the power vacuum left behind after the evaporation of governance and law enforcement, laying claim to their own personal fiefdoms. Those who seize control of the remaining resources (food, fuel, and so on) will administer the only items that have any inherent value in the new world order. Cash and credit cards will be meaningless. Those appropriating the caches of preserved food as their own "property" will become very wealthy and powerful—the new kings—controlling the allocation of food to buy loyalty and services in the same way that ancient Mesopotamian emperors did. In this environment, people with special skills, such as doctors and nurses, might do well to keep this to themselves, as they may be forced to serve the gangs as highly specialized slaves.

Lethal force may be applied swiftly to deter looters and raids from rival gangs, and as resources become depleted the competition will get only fiercer. A common mantra of people who actively prepare for the apocalypse (called Preppers) is: "It is better to have a gun and not need it than to need a gun and not have it."

One pattern likely to recur over the weeks and months after the Fall is that small communities of people will gather together in a defensible location for mutual support and protection of their own stash of consumables, looking for safety in numbers. These small dominions will need to patrol and protect their own borders in the way that whole nations do today. Ironically, the safest place for a group to barricade themselves in and hunker down during the turbulence would be one of the fortresses dotted across the country, but now turned inside out in its purpose. Prisons are largely self-contained compounds with high walls, sturdy gates, barbed wire, and watchtowers, originally intended to prevent the inhabitants from escaping, but equally effective as a defensive refuge for keeping others out.

The outbreak of widespread crime and violence is probably an inevitable effect of any catastrophic event. However, this hellish descent into a *Lord of the Flies* world is not something I will discuss any further here. This book is about how to fast-track the recovery of technological civilization once people are able to settle down again.

THE BEST WAY FOR THE WORLD TO END

Before we get to the "best" let's start with the worst. From the point of view of rebuilding civilization, the worst kind of doomsday event would be all-out nuclear war. Even if you escape vaporization in the targeted cities, much of the material of the modern world will have been obliterated, and the dust-darkened skies and ground poisoned by fallout would hamper the recovery of agriculture. Just as bad, even though it is not directly lethal, would be an enormous coronal mass ejection from the Sun. A particularly violent solar burp would slam into the magnetic field around our planet, set it ringing like a bell, and induce enormous currents in the electricity distribution wires, destroying transformers and knocking out electrical grids across the planet. The global power blackout would disrupt the pumping of water and gas supplies and the refining of fuel, as well as the production of replacement transformers. With such devastation of the core infrastructure of modern civilization without any immediate loss of life, the collapse of social order would soon follow, and the roving crowds would rapidly consume the remnant supplies and so precipitate a subsequent mass depopulation. At the end, survivors would still encounter a world without people, but one that has now also been stripped bare of any resources that would have offered them a grace period for recovery.

While the dramatic scenario favored by many post-apocalyptic movies and novels may be the collapse of industrial civilization and social order, forcing survivors to engage in an increasingly frantic struggle for dwindling resources, the scenario that I want to focus on is the inverse: a sudden and extreme depopulation that leaves the material infrastructure of our technological civilization untouched. The majority of humanity has been erased, but all of the stuff is still around. This scenario presents the most interesting starting point for the thought experiment on how to accelerate the rebuilding of civilization from scratch. It grants the survivors a grace period to find their feet, preventing a degenerative slide too far, before they need to relearn all the essential functions of a self-supporting society.

In this sense, the "best" way for the world to end would be at the hands of a fast-spreading pandemic. The perfect viral storm is a contagion that combines aggressive virulence, a long incubation period, and near 100 percent mortality. This way, the agent of the apocalypse is extremely infectious between individuals, takes a little while for its sickness to kick in (so that it maximizes the pool of subsequent hosts that are infected), but results in certain death in the end. We have become a truly urban species—since 2008 more than half of the global population lives in cities rather than rural areas—and this crammed density of people, along with fervent intercontinental travel, provides the perfect conditions for the rapid transmission of contagions. If a plague like the Black Death, which wiped out a third of the European population (and probably a similar proportion across Asia) from 1347 to 1351, were to strike today, our technological civilization would be much less resilient.[1]

---

1 However, some of the longer-term ramifications of the Black Death were beneficial to society: a cultural silver lining to the cloud that was the Great Dying. With the ensuing labor shortage, serf peasants surviving the mass depopulation were able to slip their bond to the lord of the manor, helping break the oppressive feudal system and usher in a much more egalitarian social structure and market-orientated economy.

WHAT, THEN, is the minimum number of survivors of a global catastrophe that is sufficient for humanity to have a feasible chance of not just repopulating the world but being able to accelerate the rebuilding of civilization? To put it another way: What is the critical mass to enable a rapid reboot?

There are two extremes on the spectrum of surviving populations, which I will call the "Mad Max" and "I Am Legend" scenarios. If there is an implosion of the technological life support system of modern society but no immediate depopulation (such as would be triggered by a coronal mass ejection), most of the population survives to rapidly consume any remaining resources in fierce competition. This wastes the grace period, and society promptly descends into *Mad Max*–style barbarism and a subsequent mass depopulation, with little hope of rapidly bouncing back. If, on the other hand, you are the sole survivor in the world, or at least one of a small number of survivors so dispersed that they are unlikely to stumble across one another during their lives, then the notion of rebuilding civilization, or even recovering the human population, is nil. Humanity hangs on by a single thread and is inevitably doomed when this Omega man or woman dies—the situation in Richard Matheson's novel *I Am Legend*. Two survivors—a male and a female—is the mathematical minimum for continuation of the species, but the genetic diversity and long-term viability of a population growing from just two individuals would be seriously compromised.

So what is the theoretical minimum needed for repopulation? Analysis of the mitochondrial DNA sequences in the Maori people living in New Zealand today has been used to estimate the number of founding pioneers who first arrived on rafts from Eastern Polynesia. The genetic diversity revealed that the effective size of this ancestral population was no more than about seventy breeding females, and so a total population a little over twice that. A similar genetic analysis deduced a comparable founding population of the great majority of Native Americans, who are descended from ancestors who crossed the Bering land bridge from Eastern Asia 15,000 years ago when sea levels were lower. Thus a post-apocalyptic group of a few hundred men and women, all in the same place, ought to encapsulate sufficient genetic variability to repopulate the world.

The problem is that even with a growth rate of 2 percent per annum, the fastest the world's population has ever grown when sustained by industrialized agriculture and modern medicine, it would still take eight centuries for this ancestral group to recover to the population of the time of the Industrial Revolution. (We'll explore in later chapters the reasons why advanced scientific and technological developments probably require a certain population size and socioeconomic structure.) And such a diminished initial population would probably be far too small to be able to actually maintain reliable cultivation, let alone more advanced production methods, and so would regress all the way back to a hunter-gatherer lifestyle, preoccupied with the struggle for subsistence. Ninety-nine percent of human existence has been spent in this lifestyle, which cannot support dense populations and represents a trap that is very hard to progress out of again. How do you avoid regressing that far?

The surviving population would need plenty of hands to work the fields to ensure agricultural productivity, yet leave enough individuals available to work on developing other crafts and recovering technologies. For the best possible restart, you'd also want the survivors to number enough that a broad swath of skill sets is represented and sufficient collective knowledge is retained to prevent sliding backward too far. Thus an initial surviving population of around 10,000 in any one area (which for a large state such as Texas represents a survival fraction of only 0.04 percent), who are able to gather into a new community and work peacefully together, represents the ideal starting point for this thought experiment.

So let's turn our attention to the sort of world that the survivors will find themselves in, and how it will change around them as they rebuild.

RECOLONIZATION BY NATURE

Immediately after the termination of routine maintenance, nature will seize its opportunity to reclaim our urban spaces. Trash and detritus will collect on the streets and pavements, blocking drains and causing the pooling of water and accumulation of debris rotting into mulch. Pioneering weeds will first begin proliferating in pockets like this. Even in the complete absence of pounding car tires, cracks in the asphalt will steadily expand into crevices. With every frost, water pooled in these depressions will freeze and expand, crumbling the hard artificial ground from within with the same punishing freeze-thaw cycle that steadily wears down entire mountain ranges. This weathering creates more and more niches for small opportunistic weeds, and then shrubs, to become established and further break up the surface. Other plants are more aggressive, their penetrating roots pushing right through the bricks and mortar to find purchase and tap into sources of moisture. Vines will snake their way up traffic lights and street signs, treating them like metallic tree trunks, and lush coatings of creepers will grow up the cliff-like faces of buildings and spread down from the rooftops.

Over a number of years, accumulating leaf litter and other vegetative matter from this pioneering burst of growth will decay to an organic humus and will mix with the windblown dust and grit of deteriorating concrete and bricks to create a genuine urban soil. Papers and other detritus billowing out of broken office windows will collect in the streets below and add to this composting layer. A thickening carpet of dirt will smother the roads, sidewalks, parking lots, and open

BUILDINGS CRUMBLE AND NATURE RECLAIMS OUR URBAN SPACES, INCLUDING OUR STORES OF KNOWLEDGE LIKE THIS NEW JERSEY LIBRARY.

plazas of towns and cities, allowing a succession of larger trees to take root. Away from the asphalt streets and paved squares, the cities' grassy parks and the surrounding countryside will rapidly return to woodland. Within just a decade or two, elder thickets and birch trees will have become firmly established, maturing to dense woods of spruce, larch, and beech trees by the end of the first century after the apocalypse.

And while nature is busy reclaiming the environment, our buildings will crumble and decay among the growing forests. As vegetation returns and fills the streets with wood and drifts of windblown leaves, mingling with the trash strewn out of broken windows, piles of perfect kindling will collect in the streets, and the chances of raging urban forest fires increase. Tinder accumulated against the side of a building and ignited by a summer lightning storm, or perhaps by sunlight focused through broken glass, is all that's needed to unleash devastating wildfires that would spread along the streets and burn up the insides of high-rises.

A modern city wouldn't be razed to the ground like London in 1666 or Chicago in 1871, the fire ripping rapidly from one wooden building to the next and leaping across the narrow streets; but blazes spreading unopposed by firefighters would still be devastating. Gas lingering in underground pipes and throughout buildings would explode, any fuel left in the tanks of vehicles abandoned in the streets only adding to the intensity of the inferno. Dotted throughout populated areas are bombs waiting to go off when a blaze sweeps through: gas stations, chemical depots. Perhaps one of the most poignant sights for post-apocalyptic survivors would be watching the burning of the old cities, sprouting thick columns of choking black smoke towering above the landscape and flushing the sky bloodred at night. After a passing blaze, the brick, concrete, and steel matrix of contemporary buildings would be all that is left behind—charred skeletons after their combustible internal viscera have been gutted.

Fire will wreak devastation across great areas of the deserted cities, but it is water that will eventually bring certain destruction for all our carefully constructed buildings. The first winter after the Fall will see a spate of burst frozen water pipes, which will disgorge inside buildings during the following thaw. Rain will blow in through missing or broken windows, trickle down among dislodged roofing tiles, and overflow from blocked gutters and drains. Peeled paint from window and door frames will allow moisture to soak in, rotting wood and corroding metal until the whole insert falls out of the wall. The wooden structures—floorboards, joists, and roof supports—will also soak up moisture and rot, while the bolts, screws, and nails holding the components together rust.

Concrete, bricks, and the mortar smeared between them are subject to temperature swings, soaked with water trickling down from blocked gutters, and pulverized by the relentless pulsing of freeze-thaw at high latitudes. In warmer climates, insects such as termites and woodworms will join forces with fungi to eat away at the wooden components of buildings. Before too long, wooden beams will decay and yield, causing floors to fall through and roofs to collapse, and eventually the walls themselves will bow outward, then topple. The majority of our houses or apartment blocks will last, at most, a hundred years.

Our metal bridges will corrode and weaken as the paint peels off, allowing water to seep in. The death knell for many bridges, though, is likely to be windblown detritus collecting in the expansion gaps, breathing spaces designed to allow the materials to swell in the summer heat. Once clogged, the bridge will strain against itself, shearing off corroding bolts until the whole structure gives way. Within a century or two, many bridges will have collapsed into the water below, the lines of rubble and debris at the feet of the still-standing pillars forming a series of weirs in the river.

The steel-reinforced concrete of many modern buildings is a marvelous building material, but although more resistant than wood, it is by no means impervious to decay. The ultimate cause of its deterioration is ironically the source of its great mechanical strength. The steel rebars are cocooned from the elements by the concrete surrounding them, but as mildly acidic rainwater soaks through, and humic acids released by rotting vegetation seep into the concrete foundations, the embedded steel begins to rust inside the structures. The final blow for this modern construction technique is the fact that steel expands as it rusts, rupturing the concrete from the inside, leaving even more surface exposed to moisture and so accelerating the endgame. These rebars are the weak point of modern construction—and unreinforced concrete will prove more durable in the long run: the dome of the Pantheon in Rome is still going strong after two thousand years.

The greatest threat to high-rises, though, is waterlogged foundations from unmaintained drainage, blocked sewers, or recurring floods, particularly among cities built along the banks of a river. The supports will corrode and degrade, or subside into the ground to create listing skyscrapers far more ominous than the leaning tower of Pisa, before inevitably collapsing. The raining debris will further damage surrounding edifices, or the buildings will perhaps even topple over into neighboring monoliths like giant dominoes, until only a few remain spiking above a skyline of trees. Few of our great high-rise buildings would be expected to still be standing after a few centuries.

So within just a generation or two after the Fall, the urban geography will have become unrecognizable. Opportunistic seedlings have become saplings have become full-blown trees. City streets and boulevards have been replaced by dense corridors of forest crammed into the man-made canyons between high-rise buildings, themselves now grossly dilapidated and trailing vegetation from gaping windows like vertical ecosystems. Nature has utterly reclaimed the urban jungle. Over time, the jagged piles of rubble from collapsed buildings will themselves become softened by the accumulation of decomposing plant matter forming soil—hillocks of dirt sprouting trees, until even the tumbled remains of once-soaring skyscrapers are buried and hidden by verdant growth.

Away from the cities, fleets of ghost ships will be adrift across the oceans, occasionally carried by the vagaries of wind and currents to ground themselves on a coastline, slicing open their bellies to bleed noxious slicks of fuel oil or releasing their load of containers onto the ocean currents like dandelion seeds in the wind. But perhaps the most spectacular shipwreck, if anyone happens to be in the right place at the right time to watch it, will be the return of one of humanity's most ambitious constructions.

The International Space Station is a giant 100-meter-wide edifice built over fourteen years in low Earth orbit: an impressive assemblage of pressurized modules, spindly struts, and dragonfly wings of solar panels. Although it soars 400 kilometers over our heads, the space station is not quite beyond the wispy upper reaches of the atmosphere, which exert an imperceptibly slight but unrelenting drag on the sprawling structure. This saps the space station's orbital energy so that it spirals steadily toward the ground, and it needs to be repeatedly boosted back up with rocket thrusters. With the demise of the astronauts, or lack of fuel, the space station will relentlessly drop about 2 kilometers every month. Before too long, it would be hauled down into a fiery plunge through the air, ending in a streak of light and fireball like an artificial shooting star.

THE POST-APOCALYPTIC CLIMATE

The gradual decay of our cities and towns is not the only transforming process that the survivors of the apocalypse will witness.

Since the Industrial Revolution and exploitation of first coal and then natural gas and oil, humanity has been fervently burrowing underground to dig up the buried chemical energy accumulated from times past. These fossil fuels, readily combustible dollops of carbon, are the decayed remains of ancient forests and marine organisms: chemical energy derived from the trapping of sunlight that shone on the Earth eons ago. This carbon originally came from the atmosphere, but the problem is that we are burning these stores so quickly that a few hundred million years' worth of fixed carbon have been released back into the atmosphere in just over a hundred years, pumped out of our smokestacks and car exhaust pipes. This is far, far faster than the planetary system can reabsorb the liberated carbon dioxide, and there is about 40 percent more of the gas in the air today than at the beginning of the eighteenth century. One effect of this elevated carbon dioxide level is that more of the Sun's warmth is trapped by the Earth's atmosphere through the greenhouse effect, leading to global warming.

This in turn will lead to a rise in sea levels and the disruption of weather patterns worldwide, creating more frequent, heavier monsoon floods in some areas and droughts in others, with severe repercussions for agriculture.

With the collapse of technological civilization, emissions from industry, intensive agriculture, and transport would cease overnight, and pollution from the small surviving population would drop to practically zero in the immediate aftermath. But even if emissions were to completely stop tomorrow, the world will continue to respond for the next few centuries to the vast amount of carbon dioxide our civilization has already belched out. We are currently in a lag phase, as the planet reacts to the sudden hard shove we have given to its equilibrium.

The post-apocalyptic world is therefore likely to experience a rise in sea level of several meters over the following centuries from momentum already built up in the system. The effects could be much worse if the warming triggers other secondary effects, such as the thawing of methane-laden permafrost or widespread melting of glaciers. While carbon dioxide levels will decline after the apocalypse, they will plateau at a substantially elevated value and not return to their preindustrial state for many tens of thousands of years. So over the timescale of our, or any following, civilization, this forced cranking-up of the planet's thermostat is essentially permanent, and our current carefree lifestyle will leave a long, dark legacy for those inhabiting the world we leave behind. The consequences for survivors already struggling to support themselves is that as climate and weather patterns continue to change over the generations, once-fertile cropland may be ruined by drought, low-lying regions become flooded, and tropical diseases become more prevalent. Shifts in local climate have caused abrupt collapses of civilizations in human history, and the ongoing global changes may well frustrate the recovery of a fragile post-apocalyptic society.

CHAPTER 2

THE GRACE PERIOD

Thus we never see the true state of our condition
till it is illustrated to us by its contraries, nor know
how to value what we enjoy, but by the want of it.

DANIEL DEFOE, *Robinson Crusoe* (1719)

AFTER A PLANE CRASH IN A REMOTE AREA, your main priorities for survival would be shelter, water, and food. The same requirements are paramount after the crash of civilization. While it's possible to survive several weeks without food, and a few days without drinking water, if you're caught outside in an inclement climate, you can die of exposure within a matter of hours. As the British Special Air Service (SAS) survival expert John "Lofty" Wiseman told me, "If you're still on your feet after the big bang, you are a survivor. But how long you continue to survive is down to your knowledge and what you do." For our purposes we'll assume that, like more than 99 percent of people, including myself, you're not a Prepper and have not stockpiled food and water, fortified your home, or made any other prior arrangements for the end of the world.

So during the crucial buffer period before you're forced to start producing things anew, what remnants could you scavenge to ensure your survival in the post-apocalyptic world? What would you want to look out for when beachcombing the detritus left behind by the receding technological tide?

In the situation we've imagined (loss of people, but no massive destruction of the stuff that surrounds us), you're not likely to want for shelter: there will be no shortage of abandoned buildings in the immediate aftermath. It would be well worth it, though, to embark immediately on a scavenging foray to a camping store to get yourself some new attire. The dress code for the end of the world will be pragmatic: loose, durable trousers, layers of warm tops, and a decent waterproof jacket will keep you comfortable while you spend a lot more time in the open or in unheated buildings. Sturdy hiking boots may not look very glamorous, but in a post-apocalyptic world you really don't want to lose your footing and break your ankle. Over the first few years, the best place to forage for clothing that has not yet been destroyed by insects or the penetrating damp would be large shopping centers. It's a long way into the deep interior of a mall, and goods would be safe from the elements.

Warm clothes aside, it is fire that will best ensure your survival. Fire has played a fundamental role in human history, protecting against the cold, providing light, enabling us to smelt metals and cook food to render it more digestible and pathogen-free. In the immediate aftermath of the collapse, you won't need wilderness survival skills such as rubbing sticks together to ignite tinder. There will be plentiful boxes of matches left in convenience stores and homes, and disposable lighters will continue working for years.

IF YOU CAN'T FIND MATCHES OR A LIGHTER, there are less conventional methods for starting a fire using scavenged materials. If it's a bright day, sunlight can be concentrated into a hot focus using a magnifying glass, a pair of eyeglasses,[1] or even the curved base of a soda can that has been polished with a square of chocolate or a dab of toothpaste. Sparks can be generated by touching together jumper cables attached to an abandoned car battery and steel wool scavenged from a kitchen cupboard will ignite spontaneously when it is rubbed against the terminals of a 9-volt battery liberated from a smoke detector. There will be an abundance of excellent tinder lying around deserted human habitations, such as cotton, wool, rags, or paper, especially if you douse it in a makeshift fire accelerant like Vaseline, hair spray, paint thinner, or simply a drop of gasoline. And you won't struggle to find fuel to burn, even in an urban environment. Populated areas are packed with combustible materials, from furniture and wooden fittings to garden shrubs, that can be thrown on a fire for heat and cooking.

The issue is not starting a fire or keeping it going, but where to make it. The vast majority of recently built houses and apartments have no working fireplace. If need be, you can safely contain a fire within a metal trash can or bring a barbecue indoors, or if the apartment has a concrete floor, you could rip away a patch of carpet and light a fire directly on the concrete. You'll need to allow the smoke and fumes to escape through a slightly opened window (especially if you're forced to resort to combusting synthetic fabrics or furniture foam). But your best bet would be to try to find an older cottage or farmhouse that is appropriately equipped to be heated by fires rather than radiators—this is one of the major incentives to abandon the cities as quickly as possible after the Fall, as we'll see in a bit.

---

1 Although only those for correcting farsightedness: the concave lenses for nearsightedness, which affects most people, disperse the light rays rather than focusing them. William Golding famously made this mistake in *Lord of the Flies*, with the nearsighted Piggy using his spectacles to start fires.

WATER

After shelter and protection from the elements, the second priority on your list is to secure clean drinking water. Before the municipal water supply runs dry you should fill your bathtub and sinks to the brim with water, as well as any clean buckets or even strong polyethylene trash bags. These emergency water stores should be covered to keep them free from detritus and to block the light that allows algal growth. Bottled water can be scavenged from supermarkets and from water coolers in office buildings. Other reservoirs of water you'll be able to drain include hotel and gym swimming pools, as well as the hot water tanks in any large building. In time, you'll come to rely on water sources you'd normally have wrinkled your nose at. Every survivor will need at least three liters of clean water every day, and more in hot climates or with exertion. And keep in mind that this is for rehydration alone, and does not include water necessary for cooking and washing.

Water that doesn't come from a sealed bottle must be purified. A surefire method for sterilizing water of pathogens is to bring it to a hearty boil for a few minutes (although this offers no protection against chemical contamination). This is very time-consuming, however, and will rapidly eat through stocks of fuel. A more practical, longer-term solution for purifying larger volumes of water, once you have settled down after the Event, relies on a combination of filtration and disinfection. A rudimentary but perfectly adequate system for filtering out particles in murky lake or river water uses a tall receptacle such as a plastic bucket, a steel drum, or even a well-cleaned trash can. Punch some small holes in the bottom, and cover with a layer of charcoal, either taken from a hardware store or created yourself using the instructions on page 107. Alternate layers of fine sand and gravel on top of the charcoal. Pour the water into your receptacle, and as it drains through, it will be effectively filtered of most particulate matter.

The first option for disinfecting this filtered water to eliminate waterborne pathogens is to use dedicated water-purification treatments, such as iodine tablets or crystals available from camping stores. If you can't find any, there are some surprising alternatives that will also work perfectly well, such as chlorine-based bleaches formulated for household cleaning. Just a few drops of a 5 percent liquid bleach solution that has sodium hypochlorite listed as the main active ingredient will disinfect a whole liter of water in an hour. But carefully check the label to ensure that the product doesn't also contain additives such as perfumes or colorants that may be poisonous. Several fluid ounces of bleach found under a kitchen sink can purify around 500 gallons of water—almost two years' supply for one person.

Products used for chlorinating swimming pools, scavenged from the storeroom of a gym or wholesaler, can also be used at a weaker dilution to disinfect drinking water. A single teaspoon of this calcium hypochlorite powder is enough to disinfect 200 gallons of water (but again, be careful it doesn't contain any antifungal agents or clarifier additives). Later on in the rebooting process, once all the readily available chlorinating agents are gone, you'll need to create your own from scratch using seawater and chalk as raw materials, as we'll see in Chapter 11.

Plastic bottles can be used not just for storing water, but for sterilizing it as well. Solar water disinfection, or SODIS, employs only sunlight and transparent bottles, and is recommended by the World Health Organization for decentralized water treatment in developing nations—a perfect low-tech option for the post-apocalyptic world. Tear the labels off clear plastic bottles—but don't use bottles bigger than two liters, as the crucial part of the Sun's rays won't be able to penetrate all the way through—fill them with the water to be disinfected, and lay them down outside in full sunlight. The ultraviolet component of the Sun's rays is very damaging to microorganisms, and if the water warms up to above 50 degrees Centigrade (122°F), this deactivating effect is greatly enhanced. A good system is leaning a sheet of corrugated iron angled to the Sun and stacking the water bottles in the grooves. Painting the sheet black helps the heat sterilization effect.

However, glass and some plastics, such as PVC, block out the UV rays. Check the bottom of the plastic bottle: most are now manufactured with a recycling symbol, and you want to pick out those marked with a (1), indicating they are made of PET. If the water is too murky for the sunlight to penetrate, you'll need to filter it first. In bright, direct sun, this method can disinfect water in around six hours, but if the sky is cloudy it's best to leave it for a couple of days.

FOOD

How long will you be able to continue dining out on the leftovers of our civilization? The expiration date on modern packaging is only a guideline and often underestimates deterioration by a considerable margin. So how long would different food types actually remain edible? Some products will last more or less indefinitely, including salt, soy sauce, vinegar, and sugar (as long as it stays dry), and we'll see in Chapter 4 how these substances can be used to preserve food.

Other staples of our diet won't fare as well on the shelves of deserted supermarkets. Most of the fresh fruit and vegetables will have wilted and rotted within weeks, but tubers will persist much longer, since they evolved to store energy over winter for the plant. Potatoes, cassava, and yams will all have a good chance of lasting more than six months if they're in a cool, dry, and dark place.

Cheese and other treats on the delicatessen counter will be moldy within weeks, and after a matter of months the butcher's unpackaged meat cuts will have decomposed to leave only the odd T-bone or rack of ribs. Eggs are actually surprisingly resilient and can remain edible for more than a month without refrigeration.

Fresh milk will be spoiled within a week or so, but "shelf safe" milk in UHT packs (pasteurized at ultra-high temperatures) will last years, and powdered milk even longer. Since it's the fat content of dried food-stuffs that often spoils first as it undergoes rancidification, fat-free powdered milk will remain potable the longest. Lard and butter will spoil quickly within defunct refrigerators, and cooking oils will also turn rancid over time. (But once unfit for human consumption, their lipid content can still be used to make soap or biodiesel, as we'll see later on.)

White wheat flour will keep for only a few years, but longer than whole wheat flour, which, due to the much higher oil content, goes rancid quickly. Flour products such as dried pasta will also last for a few years. The nutritional content survives far better if the grains have not been cracked or ground (which exposes the inner germ to moisture and oxygen), so unmilled whole wheat grains remain good for decades. Likewise, whole corn kernels will remain nutritious for around ten years, but this persistence time drops to only two or three years for cornmeal. Dried rice will keep well for between five and ten years.

This all assumes that the remnant food will be in conditions favoring preservation: cool and dry. This isn't an unreasonable expectation for the interior of a large supermarket in temperate regions, but if you're living in a hot, humid climate, food will begin to decay rapidly as soon as the grid goes down and air conditioners rumble to silence. After the refrigerators and freezers fail, the pungent aroma of putrefying food will attract many nonhuman foragers: rats and insects, as well as packs of dogs and other former pets now growing increasingly hungry. Even well-packaged food is likely to succumb to the onslaught of teeth and claws, so the food resources available to the survivors may be limited less by expiration dates than by pests—no different from the granaries of the earliest civilizations.

By far the largest reserve of preserved sustenance, however, will be the rows upon rows of canned food that fill the supermarket shelves. The armored packaging will not only resist the post-apocalyptic plagues of vermin and insects, but the heat treatment during the canning process is exceptionally good at protecting their contents against microbial spoilage from within. Although the printed "best before" date is often only two years in the future, many canned products will keep for several decades, if not more than a century after the fall of the civilization that produced them. Rust or dents on the can itself do not necessarily mean that the contents are compromised, as long as it shows no signs of leakage or bulging.

So if you were a survivor with an entire supermarket all to yourself, for how long could you subsist on its contents? Your best strategy would be to consume perishable goods for the first few weeks, and then turn to the dried pasta and rice, as well as the more resilient tuber vegetables, before finally resorting to the most reliable reserve of canned produce. Assuming also that you are careful to keep a balanced diet with the necessary intake of vitamins and fiber (the health supplements aisle will help you here), your body will need 2,000 to 3,000 calories a day, depending on your size, gender, and how active you are. A single average-size supermarket should be able to sustain you for around 55 years—63 if you eat the canned cat and dog food as well.

This calculation naturally scales up, from a single individual with a supermarket at his or her disposal to the surviving population of a cataclysm surrounded by the preserved sustenance of a whole nation, from small corner stores to enormous distribution warehouses. For example, the UK Department for Environment, Food and Rural Affairs (DEFRA) estimated in 2010 that Great Britain has a national stock reserve of 11.8 days of "ambient slow-moving groceries" (nonperishable, unfrozen produce such as rice, dried pasta, and canned goods). With an apocalyptic population crash, this would equate to up to fifty years supply for a surviving community of a few tens of thousands of people.

Thus a post-apocalyptic community large enough to rapidly reboot technological civilization should have sufficient breathing room to reinstate agriculture and grow its own food.

FUEL

Another key consumable of modern life, and one that will remain crucial for transport, agriculture, and running generators during the rebuilding, is fuel. There will be huge reserves of gasoline and diesel fuel for the surviving population. The fuel tanks of millions and millions of cars, motorcycles, buses, and trucks offer a scattered repository that can be tapped into. Gasoline can be scavenged from abandoned cars by siphoning it out of the tank, or even more simply by hammering a screwdriver into the tank to drain it into a waiting receptacle. The underground storage tanks of gas stations also collectively hold a vast reserve. Without power, the gas pumps won't work, but it wouldn't take much to jury-rig a pump with a 16-foot pipe to drain it. Each gas station holds a subterranean lake of fuel of typically around 30,000 gallons, enough for an average family car to drive more than a million miles along post-apocalyptic roads.

The wider issue is how well that fuel persists. Diesel is more stable than gasoline, but even within a year, reactions with oxygen would begin to form a gummy sediment that clogs the filters in engines, and accumulated water from condensation would permit microbial growth. If well protected and filtered before use, stored fuel might still be good for a decade or so before you'd need to start finding ways to reprocess it for continued use.

Motor vehicles themselves can be kept rolling as components wear out and fail by cannibalizing replacements from other automobiles, or improvising. Cuba offers a good contemporary example of this. The 1962 US embargo abruptly isolated the island from imports of American technology or machine components. Many of the cars still on the road today are classic models, nicknamed Yank Tanks, dating back to before this cutoff. The only reason these vehicles are still working fifty years later is the ingenuity of Cuban mechanics, who improvise repairs or harvest replacement components from other cars "parted out." These repairmen are forced to be increasingly ingenious as the pool of working parts steadily diminishes: a pattern that will certainly be repeated on a larger scale during the grace period following the collapse of civilization.

While fuel stocks and cannibalized parts will keep cars, planes, and boats going for a while, the modern GPS navigation devices we have become so reliant upon will malfunction surprisingly quickly after satellites lose the regular uplink from their command center. Positional accuracy will drop to about half a kilometer within two weeks of the Fall and around 10 kilometers within six months, and the system will be utterly useless within just a few years as the satellites drift out of their precisely coordinated orbits.

MEDICINE

Medical supplies will be yet another crucial foraging target in the aftermath. Ensuring access to classes of pharmaceuticals such as analgesics, anti-inflammatories, antidiarrheals, and antibiotics will help keep you and your companions comfortable and healthy. Deserted hospitals, clinics, and pharmacies are not the only repositories of vital drugs—you should also look in pet shops and vet practices. Antibiotics marketed for farm and pet animals, and even for fish aquariums, are exactly the same as for humans and should not be overlooked in your scavenging.

Other everyday items will also be worth gathering, as they can be reappropriated for medical uses. One of the earliest uses of cyanoacrylate adhesive, better known as superglue, was for rapidly closing wounds of US soldiers during the Vietnam War. This application would become very important again in preventing life-threatening infections in a post-apocalyptic world if you don't have immediate access to sterilized suturing needles and threads. The technique is to first thoroughly wash out the wound and cleanse it with antiseptic, perhaps purified ethanol that you have distilled yourself (see page 91). Then pull the lips of the injury together and administer the superglue only along the surface to bridge the gap and hold it closed.

Your main concern, however, will be how long a stash of medications would last before they expire. In the early 1980s, the US Department of Defense found itself sitting on a $1 billion stockpile of drugs that were about to exceed the printed expiration date, and facing the prospect of having to replace that reserve every two to three years. They commissioned a study by the Food and Drug Administration to test more than a hundred different medicines to see how long each one remained effective. They found that, astonishingly, about 90 percent of the drugs tested were still effective beyond their supposed expiration date, and in many cases their actual persistence was substantially longer. The antibiotic ciprofloxacin was still good after a decade. A more recent study found that the antiviral drugs amantadine and rimantadine remained stable after twenty-five years of storage, and theophylline tablets, prescribed for respiratory diseases like COPD and asthma, still exhibited 90 percent stability more than three decades later. On the whole, it is estimated that most drugs will still be largely effective several years beyond the expiration date given by the pharmaceutical company, even if the sealed packaging has been opened. And with modern blister packs, which protect each individual pill from degradation by moisture and oxidation from the air until the moment it is needed, the persistence could be substantially longer. So if you're facing a potentially life-threatening infection, you should almost certainly take your chances with a long-expired pack of antibiotics. Although the potency of a pharmaceutical will decline as the active ingredient in the tablet chemically degrades, there's no great risk that it'll harm you.

WHY YOU SHOULD LEAVE THE CITIES

You might think that the worst thing about any city is the other people: dense swarms pouring along the streets and jostling one another onto the subway, all immersed in the roaring soundscape of traffic, car horns, and sirens. After a catastrophic depopulation the silent tranquility of a deserted metropolis would be pretty eerie at first, but might become very pleasant. Yet while the dead cities will be phenomenal resources for scavenging the materials needed for rebuilding, it's unlikely you'll be able to continue living there after the Fall.

In the immediate aftermath, the major problem with built-up areas will be the huge numbers of bodies of those who died in the catastrophe. With no organized service to remove and dispose of corpses in a sanitary way, not only will the stench of decay be unbearable for the first months, but the rot and decomposition will pose a severe health hazard. As with any disaster, transmissible diseases from contaminated water supplies will be a big concern.

But after a year or so of touring the countryside and looking for other survivors, why not move back into town with all its amenities? The fact is that the glittering skyscrapers of modern cities, and even modestly tall apartment buildings, will become practically uninhabitable with the fall of civilization: they function only with the support of modern infrastructure. With no electricity grid or natural-gas supply to run the air-conditioning units or heating system, you'd find the indoor climate uncomfortable and difficult to control. Water mains will have lost pressure, so you'd need to find some source of groundwater in the city and carry several gallons a day back to your apartment, schlepping it up the staircase with no electricity to run the elevators. With enough determination you could fix many of these inconveniences by rigging diesel generators to run the elevators, air-conditioning units, and water pumps, for example, at least for the time being. You might even briefly entertain some fantasy of moving into a plush penthouse apartment, surveying the serene, deserted city around you through its floor-to-ceiling plate-glass windows, and cultivating all you need to eat in a dense permaculture in the roof garden. A more plausible model for post-apocalyptic city dwelling would be to live immediately adjacent to a major park and plow up the turf to cultivate crops.

In some cities, the environment will quickly become uninhabitable once the technological bubble bursts. Places like Los Angeles and Las Vegas have been incongruously built in very arid or even desert locales, and will rapidly wither as maintenance fails on the aqueducts supplying them with water from afar. Washington, DC, on the other hand, will face the opposite problem, as it was built on former swampland that will begin to revert to its original state with the loss of drainage.

I suspect, therefore, that you'll find it far easier to leave the cities for good and move to a more appropriate site: a rural location with fertile, cultivable ground and older buildings better suited for off-grid habitation. The sort of location that would be good for settling again would be coastal—although be mindful of the inevitable sea-level rises due to continuing climate change—allowing access to sea fishing, and near woodland. As we will see, trees have an enormous number of different uses, not just as firewood or timber for construction. You'll be able to send foraging parties and salvage crews into the dead cities, but you'll find it much easier living in the countryside. And once you've resettled, you'll want to resurrect basic technological infrastructure as far as possible, beginning with a localized electricity network.

OFF-GRID ELECTRICITY

Unlike food or fuel, electricity cannot be stockpiled—it is provided as a continual flow, and so will disappear when the grid goes down within a matter of days after the apocalypse. To retain an electricity supply, the community of survivors will need to generate their own, and we can learn a lot about what is needed by looking at those choosing to live in a self-sustaining "off-grid" way today.

The simplest short-term solution will be to scavenge a bunch of mobile diesel-powered generators from roadwork or construction sites. You may also be able to jack in to any tall wind turbines dotted along nearby hills to keep a renewable power grid going as fuel runs out. Just one of these can provide over a megawatt of power, enough for around a thousand modern homes, until it requires maintenance that you are unable to perform without dedicated equipment or precision spare parts.

Mechanically minded survivors shouldn't have too much trouble cobbling together rudimentary windmills from salvaged materials. Thin steel sheets could be cut and curved into the radial vanes of a large fan, mounted on the hub of a wheel, and the torque could be transferred by a chain and bicycle gear set.

The principal step is to convert that rotational energy into electricity, and for that you'll want to salvage a suitable ready-made generator. A source of particularly handy and compact versions is so ubiquitous in the modern world that you would be forgiven for overlooking them. There are around a billion motor vehicles on the planet now—with the US having the most of any nation, around a quarter of the total—and each of them has a salvageable alternator. The car alternator is an ingenious mechanism. Spin the shaft and a perfectly steady 12 volts of direct current appears across its terminals, regardless of how quickly the shaft is turned, making it perfectly suited to be repurposed for post-apocalyptic small-scale power generation. Simpler alternatives would be to salvage the permanent magnet motors from power tools such as cordless drills, or from the treadmills in gyms. If you forcibly spin the motor spindle it will work backward to generate an electrical current through its terminals, although the output will vary with speed.

Solar panels can also be salvaged and, unlike a diesel generator or wind turbine, have no moving parts and so survive remarkably well without maintenance. The panels do deteriorate over time, though, from moisture penetrating the casing or sunlight degrading the high-purity silicon layers. The electricity generated by a solar panel declines by about 1 percent every year, and so after two or three generations of survivors the panels will have degraded to the point of being useless.

Storing this generated electrical energy for use is your next problem. In fact, one of the first places you'll probably want to head to after the apocalypse is the golf course, not for a relaxing 18-hole round to help ease the stress of the end of the world as we know it, but to gather a crucial resource. Car batteries are very reliable, but are designed to give a high-current, brief burst of power to spin the starter motor. They're poorly suited to providing the sustained, steady supply of electrical energy that you would need for powering your new off-grid life; in fact, they are easily damaged if persistently allowed to discharge by more than about 5 percent.

An alternative design of rechargeable lead-acid battery, known as a deep cycle, discharges at a much slower rate and can have almost its entire capacity repeatedly drained and recharged without problems. It's this kind of battery that you want to forage for in the immediate aftermath. Try caravans and other RVs, motorized wheelchairs, electric forklift trucks, and golf carts—hence the recommended trip to the course. The direct-current output from your bank of storage batteries can run many appliances, such as small fridges and lamps, but you'll

INHABITANTS OF GORAŽDE, CUT OFF FROM THE GRID BY SERBIAN FORCES IN THE MID-1990S, JURY-RIGGED RUDIMENTARY HYDROELECTRIC GENERATORS TETHERED TO A BRIDGE.

also want to recover a device called an inverter that will convert that DC into a 120-volt alternating current suitable for powering other appliances.

Electricity generation and storage setups like this are used today by Off-Gridders and Preppers steeling themselves for the collapse of civilization. But recent history has also shown us some compelling examples of ingenuity from everyday urbanites in maintaining an electrical supply during adversity. For example, during the Bosnian war of the mid-1990s, the city of Goražde was surrounded and besieged for three years by the Serbian army, and was forced to become largely self-sufficient. Although the inhabitants received airlifted UN food supplies, much of their modern infrastructure was destroyed, and they were cut off from the power grid. To generate electricity, Goraždeans built their own makeshift hydropower installations: platforms floating in the Drina River moored to the city bridges, fitted with paddle water-wheels driving scavenged car alternators. These were strangely reminiscent of the flour-grinding ship mills in medieval European cities, moored off bridges in the fastest current in the middle of the river, but the modern innovations were feeding electricity back to the riverbanks along suspended cables.

CANNIBALIZING THE CITIES

We've looked so far at how the remnants left behind by our civilization will cushion the decline of the surviving society, offering a buffer zone of commodities like food and fuel, as well as components such as alternators and batteries that can be jury-rigged into post-apocalyptic power generation. But the dead cities will also provide the basic raw materials needed for rebuilding afresh.

Some crucial materials, like many metals and glass, are easy to recycle. Even if metallic components have been badly rusted and corroded over a long period, the metal is still there. It just needs to be separated from the other elements that have become bonded with it, mostly oxygen. A heavily rusted steel girder is essentially a very rich iron ore, and can be refined back to the pure metal using the same techniques used historically to smelt iron from rocky natural ores, as we'll see later in the book.

Plastics are synthesized using sophisticated organic chemistry (and oil-derived feedstocks) and so will be available in the early stages of the recovery only by repurposing or recycling that which already exists. They fall into two camps, depending on their polymer structure and thus their response to heat: thermosetting plastics and thermosoftening plastics (or simply thermoplastics). Thermosetting plastics are nigh on impossible to recycle: when heated they decompose into a complex mixture of different organic compounds, many pretty noxious. Thermoplastics, however, once cleaned, can be melted and re-formed into new products. The easiest thermoplastic to recycle with rudimentary methods is polyethylene terephthalate (PET). The simple way to tell which specific polymer your scavenged plastic items are made from is to check the recycling identification code imprinted on them. PET is distinguished with (1)—plastic drinking bottles, for example, are almost exclusively PET—and you may also have some success recycling (2), high-density polyethylene (HDPE), and (3), polyvinyl chloride (PVC). However, while glass can be melted and re-formed indefinitely, the quality of plastic products degrades with exposure to sunlight and

the oxygen in the air, and they become weaker and more brittle each time they are recycled.[2] So while a post-apocalyptic society would be able to feed on our carcass of metal and glass, the age of plastics will inevitably draw to an end, until sufficient chemical proficiency can be relearned.

With the fall of civilization and the collapse of long-distance communication networks and air travel, the global village will shatter back into a globe of villages. The Internet, despite being originally designed as a resilient computer network to survive nuclear attack and the loss of many of its nodes, will fare no better than any other modern technology with systemic failure of the electricity grids. Mobile phones will also last only a matter of days after grid-down, once the backup generators at the computer centers and cell towers run out of

---

2 Also, modern packaging and articles are rarely formed from a single plastic type. For example, a toothpaste tube is actually composed of five layers, all extruded at the same time: linear low-density polyethylene, modified low-density polyethylene, ethyl vinyl alcohol, modified low-density polyethylene, and finally linear low-density polyethylene (fittingly, the plastic tube is itself extruded out of a nozzle, much like the toothpaste it will be filled with). This makes the plastic of many products practically unrecoverable, and so only simple articles, such as a PET clear water bottle, would be worth salvaging in a post-apocalyptic world.

fuel. This will mean that marginal or old technologies will suddenly assume a profound new importance. One of the first things you'll want to find are old-fashioned walkie-talkies for keeping in contact with other members of your group when you're separated while out scavenging. For long-range communication, citizens band or ham radio sets will become pretty valuable for trying to establish contact with other pockets of survivors.

But the most valuable resource to gather before it is lost is knowledge. Books may have been destroyed in unchecked fires ripping though cities and towns, turned to illegible mush by the pulse of floodwaters, or simply rotted on the shelves from the humidity and rain blowing in through broken windows. Although far more expansive, our civilization's paper-based writings are actually less-permanent records than the clay tablets, tough papyrus rolls, or animal-skin parchment of earlier cultures. But if the contents of libraries are still intact when the surviving population begins to rebuild, these fabulous resources can be mined for knowledge. Many of the titles listed in the bibliography at the back of this book, for example, offer details of key practical skills and the processes required for civilization, and would be well worth searching for. Equally, it would be worth trying repositories of older technology—science and industry museums—for contraptions like spinning machines or steam engines that could be studied and reverse engineered as appropriate technologies for the post-apocalyptic world.

So a scene that is likely to become common during the recovery would be growing settlements of survivors dotted across the countryside. These are not located haphazardly but arranged in rings around the dead cities, circling the core of dilapidated high-rises and other urban infrastructure. Only salvage crews venture into these uninhabited zones, picking over the bones of the dead cities, mining their remnants for the most useful materials, perhaps using homemade explosives to fell buildings and makeshift acetylene torches to dissect metal components. The valuable swag is then hauled back to be reprocessed into tools, plowshares, or whatever is needed during the reboot.

One of your earliest challenges after the Fall will be restarting agriculture. While there will be plenty of empty buildings to provide shelter, and underground lakes of fuel for running vehicles and powering generators, all will be for naught if you starve to death.

CHAPTER 3

AGRICULTURE

> We've been given a flying start in a new kind of
> world. We're endowed with a capital of enough of
> everything to begin with, but that isn't going to last
> forever. . . . Later we'll have to plow; still later we'll
> have to learn how to make plowshares; later than
> that we'll have to learn how to smelt the iron to
> make the shares. . . . The most valuable part of
> our flying start is knowledge. That's the short cut
> to save us starting where our ancestors did.
>
> JOHN WYNDHAM, *The Day of the Triffids* (1951)

THE URGENCY WITH WHICH you will need to reboot agriculture depends entirely on how many people survived whatever event precipitated the collapse of society. For the purposes of our thought experiment, you will have breathing room before the stocks of preserved food are depleted. This will give you time to find your feet, scout for some suitable land to resettle on, and gradually learn from your mistakes in the fields before a reliable harvest becomes a matter of life or death.

You'll need to move quickly after the Fall to recover and preserve as many crop plants as you can. Every modern strain of crop represents thousands of years of diligent selective breeding, and if you lose ready-domesticated species, you may lose any hope of shortcutting the re-building of civilization. Over the course of their domestication, species such as wheat and maize have been optimized for the production of nutrition and are now poorly adapted for life without us. Many will be quickly outcompeted and may be driven to extinction by wild plants seizing their opportunity to reclaim the abandoned fields.

Overgrown, abandoned fields or backyard vegetable patches would be sensible places to look for surviving edible plants. Varieties like rhubarb, potato, and artichoke are likely to keep self-propagating long after the patch is abandoned. But the staple of our diets is cereal crops, and if you were particularly conscientious, you might try to organize sampling expeditions immediately after the Fall to collect seed before the plants die and rot in the fields. Or you may be lucky enough to scavenge from abandoned farm barns sacks of seed corn still viable years later.

The problem you will face, though, is that many of the crops cultivated in modern agriculture are "hybrids": they are produced by crossing two inbred strains possessing desirable characteristics to produce progeny that are very uniform and extremely high-yielding. Unfortunately, seeds produced in turn by this hybrid crop will not retain that consistency—they do not "breed true," and so new hybrid seeds must be bought to plant each year. What you really want to gather in the immediate aftermath, therefore, are heirloom crops: traditional varieties that can be reliably propagated from one year to the next. Many Preppers stockpile heirloom seeds for exactly this eventuality, but where should you turn if you've not prepared a reserve ahead of time?

Hundreds of seed banks exist around the world, safeguarding biological diversity for posterity. The largest of these is the Millennium Seed Bank in West Sussex, just outside London. Here, billions of seeds are stored in a nuclear-bombproof, multistory underground vault, offering a vital post-apocalyptic library, not of knowledge-laden books, but of diverse crop strains. The seeds of many kinds of plants remain viable for decades in a cool, dry environment, including cereal crops, and peas and other legumes, as well as potato, eggplant, and tomato.

Svalbard Global Seed Vault

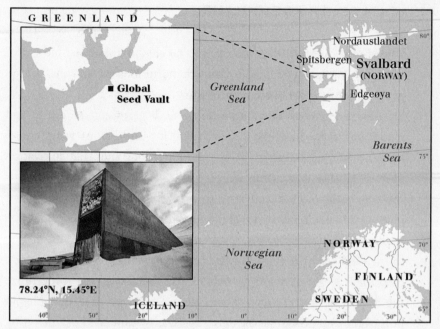

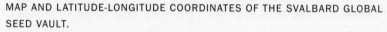

MAP AND LATITUDE-LONGITUDE COORDINATES OF THE SVALBARD GLOBAL
SEED VAULT.

But even these seeds die after a while if they have not been germinated, grown, and fresh seeds collected for continued storage.

Low temperatures extend this persistence period, and so perhaps the most resilient agricultural backup, a SAVE file that will last long after the collapse of civilization, is the Svalbard Global Seed Vault. This repository is built 125 meters into a mountainside on the Norwegian island of Spitsbergen. The 1-meter-thick steel-reinforced concrete walls, blast doors, and airlocks will protect the biological cache inside from the worst global cataclysm, and even with a loss of power the entombing permafrost (the site is well within the Arctic Circle) naturally maintains a subzero temperature for long-term preservation. Viable wheat and barley seeds will be safeguarded for more than a millennium.

PRINCIPLES OF AGRICULTURE

The crucial question that you need to be able to answer is: How do I walk out into a muddy field with a handful of seed and make food come out of it before the winter sets in?

This might seem like a no-brainer: seeds germinate naturally, and plants had been growing quite happily for millions of years before humans evolved. But that doesn't mean by a long shot that cultivation and agriculture come easy. While plants grow naturally, farming is grossly artificial. You are trying to cultivate one particular variety of plant in monoculture, a pure and uniform crop isolated in a field to the exclusion of all other plants. (Any other plants that do begin growing in the field are by definition weeds and are competing with your food crop for sunlight and soil nutrients.) You are also trying to optimize the density of crop plants in your field, to get as much as possible out of the land and minimize the effort and energy expended in cultivating large areas. But you need to prevent this juicy target from being overrun by insects and other pests or fungal diseases that run riot under such ideal conditions (in the same way that cities are perfect breeding grounds for human pathogens). These two factors mean that a field of crops is a highly synthetic environment, and nature is constantly pushing back at you. It takes a great deal of careful control and effort to maintain this unstable situation.

Yet you have an even more fundamental problem to overcome in agriculture. In a natural ecosystem such as a woodland, trees and underbrush plants grow by soaking up energy from sunlight, absorbing carbon from the air, and piping up a variety of mineral nutrients from the soil through their roots. These vital substances become incorporated into the leaves, stems, and roots of plants, and, when eaten, become part of an animal's body. When the animal later excretes, or dies and decays, these nutrients simply soak back into the soil whence they came. A natural ecosystem is therefore a healthy circulating economy of elements being transferred endlessly between different accounts. But the nature of farmland is fundamentally different: you are encouraging growth for the sole purpose of harvesting and removing the products for human consumption. Even if you spread much of the leftover vegetative matter back onto the fields, you've still removed the portion actually eaten, and year after year the land is steadily depleted. So the very nature of farming necessitates that you are progressively removing mineral nutrients, bleeding the soil of its vitality. And particularly with modern sewage systems—our waste is treated to kill harmful bacteria and then discharged into rivers or seas—agriculture today is an efficient pipeline for stripping nutrients from the land and flushing them into the ocean. Vegetation needs balanced nutrition just as much as the human body does, and the three major plant foods are the elements nitrogen, phosphorus, and potassium. Phosphorus is crucial for the transfer of energy, and potassium helps reduce water loss, but it is nitrogen, used in building all proteins, that is most often the limiting factor for crop yield. Unless you're extraordinarily lucky, like the ancient Egyptians in the Nile valley, where the annual floods revitalize the land with fertile silt, you need to take action to address this fundamental deficit in the balance sheets.

Modern industrialized agriculture is astonishingly successful, with an acre today producing two to four times more food than the same land provided a hundred years ago. But the only way that farms today can function, growing dense monocultures on the same land and still producing high yields year after year, is by spraying potent herbicides and pesticides to maintain an iron-fist control over the ecosystem, and by the liberal application of chemical fertilizers. The nitrogen-rich compounds provided in these artificial fertilizers are created industrially by the Haber-Bosch process, which we'll return to in Chapter 11. All of these herbicides, pesticides, and artificial fertilizers are synthesized using fossil fuels, which also power the farmyard machinery. In a sense, then, modern farming is a process that transforms oil into food—with some input from sunshine—and consumes around ten calories of fossil fuel energy for every calorie of food actually eaten. With a collapse of civilization and the disappearance of an advanced chemical industry, you'll need to relearn traditional methods. Today, organic produce is the preserve of the wealthy; in the aftermath it will be your only option.

We'll come back later in this chapter to how you can maintain soil fertility over the years. Let's start with the fundamentals of cultivating crops from the ground up.

WHAT IS SOIL?

As a farmer, you have only limited control over nature. You obviously cannot control the amount of sunlight beaming onto your fields: you can't change the climate of your region or dial in the seasons. You also can't control the rainfall, although you can regulate the moisture content of your fields by balancing irrigation and drainage. The one thing you have most control over is the soil: you can chemically enrich it with fertilizers, as we've just seen, and physically manipulate it with tools like the plow. So the most fundamental element of agriculture under a farmer's control is the soil, and that requires an understanding of what soil is, and how it supports plant growth.

All the civilizations of history owe their existence to this thin scraping of topsoil. Hunter-gatherers can support themselves by foraging in woodlands, but cities and civilization rely on the enormous productivity of cereal crops—shallow-rooted grasses that are utterly dependent on the resources provided by topsoil. The basis of all soil is disintegrated rocks that make up the crust of our planet. Rock is physically attacked by flowing water, blowing wind, and grinding glaciers, and chemically weathered by weakly acidic rainwater that dissolves a little carbon dioxide as it drops from the clouds. Depending on the degree of crumbling, this produces gravels, sands, and clays. These particles are stuck together with humus—a matrix of organic matter that helps retain moisture and minerals, and gives topsoil its dark color. Soils typically contain between 1 and 10 percent humus, although peats approach 100 percent organic matter. But, most important, soil hosts a huge and diverse population of microbial life, an invisible ecosystem that processes decaying matter and recycles nutrients for plants.

The main factor that determines the nature of a particular soil and its appropriateness for different crops is the proportion of different particle sizes: gritty sand, intermediate silt, and fine clay. It's easy to get a visual check on soil composition. Fill a glass jar one-third of the way with soil (picking out any hard clumps, stems, or leaves) and top it off almost to the brim with water. Screw on a lid, and shake vigorously until all lumps have been broken up and you have a uniform muddy soup. Let the jar stand undisturbed for a day or so, allowing time for the suspension to settle back down and the water to be nearly clear again. The different grains will have sedimented out in order of their particle size to show distinct layers or bands. The bottom band is the coarse-grained sand component of the soil, the middle layer is silt, and the very top layer holds the finest clay particles, allowing you to visually judge their proportions in the mixed soil.

The ideal kind of soil for farming is known as loam and is a balanced mixture of roughly 40 percent sand, 40 percent silt, and 20 percent clay. A sandy soil (more than two-thirds of the total) drains well and so is good for wintering cattle, as it won't get trodden into a quagmire, but minerals and fertilizers are easily washed out and require extra manure. On the other hand, a heavy clay soil (more than a third clay particles and less than half sand) is physically hard to work with plows and harrows, and will require more liming to maintain a healthy crumbly structure.

Wheat, beans, potatoes, and rapeseed (the source of canola oil) all grow superbly well in well-managed clay soils. Oats thrive in heavier, damper soils than are suitable for wheat or barley, such as the soils of Scotland created by the grinding sweep of glaciers in the last ice age. Historically, oats and potatoes have allowed people to achieve high yields and settle areas where other crops do not grow. Barley prefers lighter soils than wheat, and rye will grow in poorer, sandier soils than other cereals. Sugar beets and carrots also grow well in sandy soils.

Being lucky enough to find fertile loam soil in a well-drained region is only the start for rebooting agriculture. In order to give your crops the best chance of success, you're also going to need to physically work the ground. Tillage is the name for all the mechanical effort you need to put into loosening hard soil, controlling weeds, and preparing a receptive layer of topsoil (tilth) for sowing the seeds.

On a sufficiently small scale, you could get by with very rudimentary handheld tools. A hoe will do an admirable job of breaking up the topsoil and mixing in manure or green fertilizer (rotting vegetative matter) before the growing season, as well as chopping up weeds before sowing and at intervals as the crop grows. A simple dibber stick can be used to poke shallow holes in the ground with regular spacing to drop seeds into and rebury with your foot. But it's backbreaking, time-consuming work, and you'd have little opportunity for doing anything else. The history of agriculture over the millennia has been a story of improving designs of farm equipment to perform these essential functions more efficiently, to maximize the productivity of the land while minimizing the labor needed.

The iconic implement of agriculture is the plow, but its role has actually changed since the beginning of cultivation. In the fertile, easily cultivable soils of Mesopotamia, Egypt, and China, where agriculture was first developed, the primitive plow was little more than a sharpened log jabbed into the ground at an angle and hauled through the soil by oxen or human laborers. The intention was to gouge a shal-

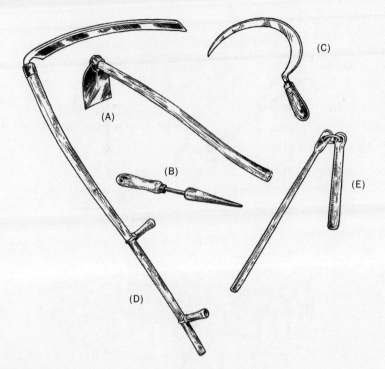

SIMPLE FARMING TOOLS: HOE (A), DIBBER (B), SICKLE (C), SCYTHE (D), THRESHING FLAIL (E).

low trench that seeds could then be dropped into and lightly buried. In most of the arable land on the planet, however, the soil needs a bit more preparation to make agriculture productive. Nowadays, the function of a plow is to carefully scoop up the uppermost layer of soil across an entire field and flip it upside down, crumbling it slightly. The primary aim of this process is weed control. Before sowing your crop on the land, undesired plants are sliced from their roots and unceremoniously covered with soil. Hidden from the sunlight, they wither and die, and their seeds are buried too deeply to successfully germinate. This cultivation of the land also helps mix organic matter and nutrients into the topsoil, particularly if you're plowing in manure, and improves drainage of the ground as well as aeration to benefit the soil microbes.

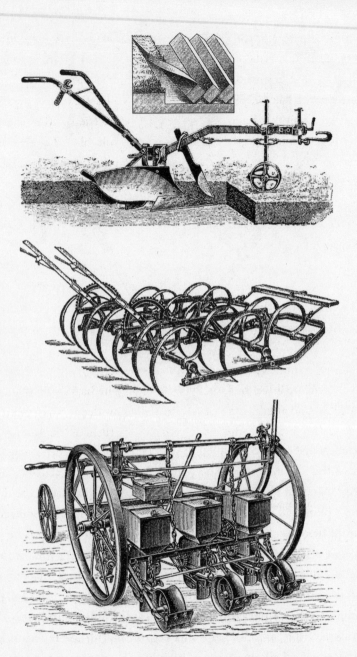

AGRICULTURAL EQUIPMENT: PLOW, HARROW, SEED DRILL. INSET: THE ACTION
OF THE PLOW TO SLICE AND TWIST OVER STRIPS OF TOPSOIL.

Immediately after the cataclysm you'll hopefully have little trouble finding abandoned tractors and fuel to run them, with trailers sporting multiple plowshares. But once available fuel dries up or lack of spare parts stills the tractor, you'll have to revert to less intensive methods. And it won't simply be a case of finding some oxen and harnessing them to the modern plow, as these large multiple-bladed contraptions require enormous traction to rip them through the ground. If you can't find a traditional plow—perhaps check museums in the nearby deserted cities—you will have to construct your own. You might be able to scavenge a modern plow blade from a trailer rack and remount it singularly on a frame, but if these have all rusted away, you could construct a wooden plow plated with cast iron, or rework scavenged steel panels in a forge. The plowshare is essentially a sharpened blade that horizontally undercuts the soil and forces it up over the moldboard, shaped to carefully roll the sod slice over and lay it back onto the field upside down.

After plowing, the resultant furrows and ridges must be smoothed down to prepare a seedbed ready for sowing. The harrow is as ancient as the plow, and alternative designs differ by how deeply they penetrate and how finely they break up clods for the tilth. Modern harrows use rows of upright metal disks to cut through the ground, or springy curved metal tines that vibrate up and down as they are dragged to pulverize the ground, mechanically mimicking the action of a hand-wielded rake. You can build your own simpler designs of diamond-shaped wooden frames with spikes sticking down, or even drag a heavy tree branch across the surface if you're really stuck. Different crops prefer particular conditions of tilth; wheat, for example, likes a fairly coarse seedbed, with clods about the size of a child's fist, whereas barley prefers a much finer tilth. Lighter harrowing is applied after sowing to cover the seeds, and can also be used between cultivated rows to tear up weeds.

Once an appropriate tilth has been prepared, the next step is to put seeds into the ground. The original meaning of "broadcast"—coined centuries before the invention of radio or TV—is scattering seeds far and wide, tossing them from a sack as you walk back and forth across the field. You can distribute seeds relatively quickly this way, but you have little control over their exact placement, which makes weeding later difficult. But again, with a little bit of ingenuity you can improve this process immeasurably. A seed drill is a mechanical seed sower. At its simplest, a cart has a hopper full of seed on top, and a chain of gears driven by one of its wheels that slowly turns a paddle at the bottom of the hopper chute to release a single seed at regular intervals. Each seed tumbles down a narrow vertical tube to be embedded at its preferred depth in the soil. Multiply the number of paddles and tubes in parallel and you can sow several rows in one pass, and tweaking the gear chain will alter the distance between plants in each row (you'll find the optimum to use for different crops by experience). This system is much less wasteful of seed because with the optimized spacing, the growing plants do not compete with one another; nor do you waste space with excessive gaps. Moreover, arraying your crops in neat lines, rather than dispersed haphazardly by broadcasting, allows you to more easily weed between the rows. With a little more sophistication, a seed drill can also be constructed to deposit a small dollop of liquid manure or fertilizer into the seed hole, thereby helping every sprout establish itself.

THE PLANTS WE EAT

Agriculture is all about exploiting a stage in the life cycle of the plants we've adopted as crops. Many plants have adapted a particular part of their structure to act as a storehouse for their captured sunlight energy, to be used either by themselves the following year or as an inheritance for the next generation, their seeds. These stores are precisely the succulent and nutritious parts that you will be familiar with on the supermarket shelves. Most of the root and stem vegetables we eat are biennials—they bloom in their second year. Their reproductive strategy is to hoard a season's worth of accumulated energy, stashed away in a specially enlarged section, remain dormant over winter, and then capitalize on their stockpile early the following spring to produce flowers and seeds well ahead of competitors. Examples of swollen taproots include carrots, turnips, rutabagas, radishes, and beets. By cultivating these strains and harvesting the bulging portions, we're essentially raiding the energetic savings account they've diligently built up over the growing season. Potatoes aren't actually a root vegetable, and the tuber we eat is in fact a swollen section of the stem. Other plants use specialized leaves as their energy storehouse—onions, leeks, garlic, and shallots are all tight clumps of thickened leaves. Cauliflower and broccoli are actually immature flowers and will become inedible if not picked early enough. Fruits are obviously the energy repository for a plant's seeds, such as the succulent flesh enveloping the pit of a plum; the grain of cereal crops such as wheat is also botanically a type of fruit.

As humanity gave up its nomadic lifestyle and became established in settlements, rooted to a particular locale with surrounding agricultural fields, we became utterly dependent on reliable harvests from the plants we adopted as crops. But we've not been content to gratefully accept the nutritious plant stores that natural selection has provided. Over many generations of selective breeding, choosing which plants to propagate on the basis of certain desirable characteristics, we've tuned their biology to emphasize certain qualities and diminish unwanted traits. In the process of hacking these plants' reproductive strategies to subvert them to our own purposes, we've distorted their biology so much that they are now as reliant upon us for their survival as we are upon them for our own. Every crop we grow today, from the gro-

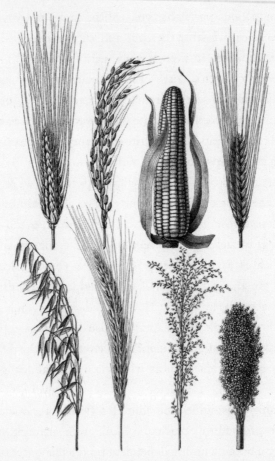

THE MOST IMPORTANT CEREAL CROPS: (TOP) WHEAT, RICE, CORN (MAIZE),
BARLEY; (BOTTOM) OATS, RYE, MILLET AND SORGHUM.

tesquely swollen tomato to the stunted and top-heavy rice plant, is a
technology in its own right, the product of ancient genetic engineers.[1]

There is an enormous diversity of edible plant species around the

---

[1] Even the familiar color of carrots is artificial: their roots are naturally white or purple,
and the orange variety was created by seventeenth-century agriculturists in the Netherlands to honor William I, the Prince of Orange.

planet, and even though only a tiny fraction of these have been chosen for cultivation and selectively bred over millennia by civilizations throughout history, there are still an estimated 7,000 cultivated strains. However, only a dozen species account for more than 80 percent of global crop production today, and the major civilizations of the Americas, Asia, and Europe were built on just three staple crops: corn, rice, and wheat, respectively. These three plants will be just as critical to rebooting after the apocalypse.

Corn, rice, and wheat, as well as barley, sorghum, millet, oats, and rye, are all cereal crops: varieties of grass. This dominance of cereals in our diets, coupled with the fact that much of the meat we consume comes from livestock fed by either grazing in pastures or on grain fodder, means that much of humanity subsists, either directly or indirectly, by eating grass. And it is on this hugely significant category of crops that survivors will need to focus.

While the harvesting of many crops is pretty straightforward and intuitive—potatoes dug out of the sod, onions plucked off the surface, and apples picked from the boughs—getting cereal grain out of the field and processed for the table is a little more involved. Harvesting maize is as simple as walking along the rows with a sack slung to your back and plucking the cobs off the stalks, but the grain of other cereals is more finicky to remove. The no-fuss method is simply to chop down the entire plant and recover the grain away from the field. The tools for reaping are the sickle and the scythe. The sickle is a short curved blade, sometimes serrated, on a handle, and is used to slice through the stalks as the other hand gathers them into bunches. The scythe is a larger, double-handed tool composed of a long pole with two grips and a gently curved blade, around a yard in length, projecting out at a right angle. Wielding a scythe takes more practice, but it is held with straightened arms, and the blade is swept horizontally over the ground in a steady rhythm by smoothly twisting the whole body. The felled stalks are tied into bundles, and these sheaves are leaned upright against one another to dry in the field, then brought inside barns before the autumn rains.

After collecting the harvest—literally reaping what you sow—the next step is to separate the grain from the rest of the plant. This is called threshing, and the simplest way is to lay the harvest out on a clean floor and beat it with a flail—a long handle with one or more shorter sticks attached at the end with leather or chain hinges. Small-scale mechanical threshers rely on exactly the same basic principle, using a spinning drum covered with pegs or wire loops that fits tightly inside a round casing to strip grain from its stalk as it passes around the gap, and sieve out the grain through a grating at the bottom.

This threshing process leaves all the grain mixed in with the empty husks, and you must now separate the wheat from the chaff (it's amazing how many everyday phrases derive from agriculture, the only vestigial link many of us retain from our heritage of working the land). The process is called winnowing, and your low-tech option is to simply toss the threshed material into the air on a windy day—the lighter chaff and straw are carried a short distance away on the breeze and the dense grain falls more or less straight back down. Modern machinery creates its own artificial wind using an electric fan, but relies on the same millennia-old principle.

As your post-apocalyptic society recovers and the population grows, one of the most crucial inventions for improving the efficiency of agriculture, producing maximum food with minimum human labor, and enabling a populous city-dwelling civilization is one that integrates these various processes. Combine harvesters today allow a single farmer to process twenty acres of wheat every hour—around a hundred times faster than reaping by hand with a scythe. A horizontal serrated blade mechanically replicates the action of the hand sickle, sawing side to side to cut the stalks as they are dragged over the front of the machine by a large cylinder of rotating paddle arms. The basic design hasn't changed in almost two centuries, and the first horse-drawn mechani-

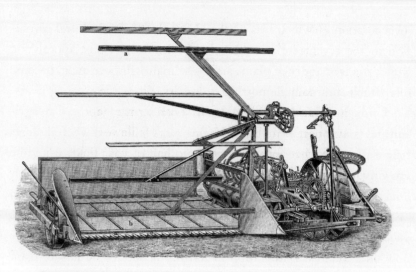

RUDIMENTARY MECHANICAL REAPER WITH SWEEPING ARMS (A) AND LOW, SICKLE-LIKE SERRATED BLADE (B).

cal reapers look surprisingly similar to their modern descendents. The combine harvester is undoubtedly one of the most important inventions of recent history, freeing so many of us from being required to work in the fields and enabling us to perform other roles in a complex society.

NORFOLK FOUR-COURSE ROTATION

As long as you can grow grain for yourself, along with some other fruits and vegetables for the sake of nutritional balance and a more interesting diet, you'll never starve to death. You could, of course, always hunt for meat, but keeping livestock, and sacrificing some of your arable capacity to support them, actually contributes a critical function for keeping your fields productive. As we've seen, without chemical fertilizers farmland would deteriorate in fertility, but animal manure allows you to return nutrients to the soil. Furthermore, there is a particular class of crops that will naturally boost soil nitrogen levels for you, the incorporation of which was a crucial step in the agricultural revolution in the seventeenth century. In the immediate post-apocalyptic world, the husbandry of plants and of animals will once again become inseparable, mutually supporting endeavors.

Throughout the Middle Ages, European farmers followed an agricultural convention of routinely leaving plots fallow—a woefully inefficient practice, as at any one time up to half of your fields would be growing no crops at all. Medieval agriculturalists recognized that their land became tired and its productivity plummeted if cereals were grown on it season after season, but they didn't understand what caused this and could only attempt a solution by resting the ground for a year. We now understand that this drop in fertility is due to the loss of plant nutrients, which is why modern agriculture is so dependent on the liberal smearing of artificial fertilizers. This solution will be closed to you for the immediate aftermath, and you'll need to revert to an older solution to the problem.

The key is that while most crops plunder nitrogen from the ground, some plants inject this vital nutrient back into the soil as they grow. This family of astounding plants is the legumes, which includes peas, beans, clover, alfalfa, lentils, soy, and peanuts. By plowing a crop of legumes back into the soil at the end of the season, or feeding them to livestock and using their manure to fertilize the land, vital nitrogen is captured and restored to the land. The incorporation of this fertility-pumping capability of legumes transformed agriculture and set Britain on course for the Industrial Revolution.

Varying between legumes and other crops on a plot of land will therefore maintain the productivity of the soil. But rather than simply swapping back and forth between two—from clover to wheat, say—a far better option is a crop rotation with several stages, as it also breaks the cycle of diseases and pests. These are often very specific to the plant they can attack, and so annually shifting, and not growing the same crop on a plot for several years, means that you can exert natural control without pesticides.

The Norfolk four-course rotation is the most successful of these historical systems and became widespread only in the eighteenth century, spearheading the British agricultural revolution. In the Norfolk system, succession of crops through each plot follows the order: legumes, wheat, root crops, barley.

As we have seen, growing legumes is intended to build up the soil's fertility for the rest of the cycle. Clover and alfalfa grow well in the British climate, but in other regions you might be better off with soy or peanuts. At the end of the season, if you're not harvesting any part of the plant for human consumption, the entire crop can be grazed by livestock or simply plowed back into the ground as green manure. The year after the legume course you want to plant a crop of wheat to capitalize on the soil fertility and produce your staple cereal for human consumption.

Don't leave the field fallow the following year, but plant a crop of a root vegetable such as turnip, rutabaga, or mangold wurzel (field beet). One of the main purposes for leaving a field fallow in the Middle Ages—to plow and harrow it in spring but leave it unplanted for a year—was to kill off weeds in preparation for the next season. But with a root vegetable, you can plant a crop and still be able to rip out weeds between the rows. This course will yield you another crop, but rather than intending all of it for your own consumption—unless the crop is potatoes—you can use it to feed the animals. Your livestock will fatten up more quickly and will also produce more manure that you can spread back onto the field to preserve its fertility. By feeding your livestock a purpose-grown fodder, rather than simply letting them forage and browse grass for themselves, you also free up pastureland, which can now be used to cultivate even more crops.

Indeed, the adoption of the humble turnip and other root crops for fodder heralded a revolution in medieval agriculture. Not only are these more effective than grazing for fattening up livestock over the summer, but they also provide a reliable energy-rich feed throughout the winter. Before their introduction, every late autumn medieval Europe witnessed the mass slaughter of livestock, as there was simply insufficient food to keep the animals from starving before spring. Turnip, as well as other fodder crops like rutabaga, kale, and kohlrabi, are biennial plants, which means they can be left in the ground over winter and plucked out to feed cattle when needed. Used to supplement the energy-poor roughage of hays and silage (fermented grass), these nutritious fodder crops support large herds of livestock through winter, continuing not only the supply of fresh meat, but also providing fresh milk and other dairy products. These are a vital source of vitamin D over the dark winter months when your skin cannot synthesize it from sunlight.

The last phase in the rotation is the planting of barley, which you can again use to feed your livestock—but remember to keep back a portion for brewing beer (as we'll cover in the next chapter). After the barley course, the rotation loops back to the beginning with the cultivation of legumes to restore the fertility of the soil and make it ready for the nitrogen-hungry cereal crop. So the rotation system is a harmonic coupling of the requirements and products of both plants and animals, it naturally combats pests and pathogens, and it allows the recycling of nutrients back into the soil. This particular system of crops won't work universally, and you'll need to find a set suited to your local soils and climate.[2] But the two key principles of the rotation system will ensure that you can reliably feed yourself and maintain soil productivity without exogenous chemical fertilizers after the apocalypse:

---

2 Even within Britain, the Norfolk four-course rotation is less effective on the heavy clay soils of the north and west, and so historically these regions focused on livestock pasturing and on manufacturing (and using their profits to buy grain from the south).

alternate legumes with cereals, and grow root crops not for your own consumption, but specifically for your livestock. Reverting to small-scale methods, five acres of land will be enough to support a group of up to ten people: wheat for bread, barley for beer, a diverse range of fruits and veggies, as well as cattle, pigs, sheep, and chickens for meat, milk, eggs, and other products.

MANURE

Spreading animal manure helps fertilize the fields, but could you exploit human waste in the same way for post-apocalyptic farming? The challenge of agriculture without modern artificial fertilizers is how to turn feces back into food (crap into crop) as efficiently as possible, and ideally you would be able to close the loop on human consumption and ensure that precious nitrogen is not lost.

At the time when the open gutters in the streets of European cities were overflowing, Chinese cities were diligently collecting their waste, not with underground sewage pipes but with buckets and carts, and spreading it on surrounding fields. Each of us produces roughly 100 pounds of feces, and around ten times as much urine, every year— waste that contains enough nitrogen, phosphorus, and potassium to fertilize crops to yield around 450 pounds of cereals.

The trouble is that you can't start gleefully smearing untreated sewage across crops you intend to eat later: you'll simply complete the life cycle of numerous human pathogens and trigger widespread outbreaks of disease. Indeed, although preindustrial China enjoyed productive agriculture, gastrointestinal diseases were endemic among the population. The proper treatment of human waste is of such crucial importance in ensuring a healthy society that you'll need to consider it right from the start as you begin to rebuild civilization. (At the very least, a post-apocalyptic settlement could dig privy pits, which should be sited at least 20 meters away from any well or stream that anyone uses as a source of drinking water.)

Disease-causing microbes and parasite eggs can be killed by heating above 65°C, or 150°F (a theme we'll come back to in the context of food preservation and health), so if you want to fertilize the fields using human manure, the problem to be solved now becomes: How do you pasteurize large volumes of your own excrement?

On a small scale, feces can be treated by sprinkling with sawdust, straw, or other non-leafy vegetative matter (to rebalance the carbon and nitrogen levels, as well as soak up moisture) before piling them in a regularly turned compost heap for several months to a year. As bacteria partly decompose the organic matter in the compost they release heat (just as our bodies' metabolism does), and this can naturally raise the heap's temperature enough to kill troublesome microorganisms. It's also best to separate urine and feces—practically achievable by simply building toilets with a funnel toward the front—to avoid a waterlogged sludge. Urine is sterile and so can be diluted and applied directly to the land.

But with a little more ingenuity, some of the human and farmyard waste can be turned into something altogether more useful with a bioreactor. In a compost heap the objective is to keep everything well aerated so that oxygen-needing bacteria and fungi can readily decompose the matter. But if instead you hold the waste in a closed vessel, blocking oxygen from getting in, anaerobic bacteria thrive and partly convert the organic material into flammable methane gas. This can be piped into a simple gas storage facility constructed from a concrete-lined pool filled with water with an upturned metal container fitted snugly within it. As the methane bubbles up into the storage tank, the water forms an air seal, and the metal gas collector rises. The weight of the floating storage tank provides gas pressure, and the methane can be piped off to supply stoves, gas for lighting, or even, as we'll see later, fuel for vehicle engines. A metric ton of organic waste can produce at least 50 cubic meters of flammable gas, equivalent to the energy of a full tank of gasoline. (It is not surprising that such biogas digesters became common across fuel-starved Nazi-occupied Europe during the Second World War.) The microbial growth slows considerably at lower temperatures, so it's important to keep the bioreactor insulated, or even siphon off some of the methane produced to heat it.

As the population of the post-apocalyptic society begins to grow again, larger-scale methods for dealing with waste will be required. Enteric bacteria, including potentially pathogenic strains, thrive in the warm internal conditions of the human body, but are poorly adapted for rapid growth outside. So the principal trick of sewage treatment is to force human enteric bacteria to compete with environmental microorganisms in a pool of poo—a survival struggle they will lose. Modern treatment plants accelerate this process by bubbling air through the sludge to encourage oxygen-needing bugs.

Although fertilizing fields with human waste may seem anathema to many of us in the Western world, it is proving to be very effective in some places. In Bangalore, India's third largest city with around eight and a half million inhabitants, euphemistically named "honey-sucker" trucks empty urban septic tanks and transport their load to surrounding agricultural areas. The waste is treated in pools before being spread on the fields. There are even commercially available products that contain processed human sewage sludge. Dillo Dirt, a fertilizer sold by the City of Austin, Texas, uses a composting process to ensure waste is naturally heated to pasteurizing temperatures to eliminate pathogens.

Aside from nitrogen, plants also need phosphorus and potassium. Bones are very rich in phosphorus—together with teeth, they are biological deposits of the mineral calcium phosphate—and so sprinkling bone meal, which is just boiled and crushed animal skeletons, is another good way of restoring failing land. Reacting the bone meal with sulfuric acid (see Chapter 5 on how to produce this) makes the phosphate much more absorbable for plants and so produces a far more effective fertilizer. In fact, the first fertilizer factory in the world was set up in 1841 to react sulfuric acid from London's gasworks with bone meal from the city's abattoirs and sell the "superphosphate" granules to farmers. Potassium for fertilizers is present in potash, which we will see in Chapter 5 is easy to extract from wood ashes; in 1870 the vast forests of Canada were the main source for fertilizers for Europe. Today we gather potassium and phosphorus for fertilizers from particular rock and mineral deposits, and identifying these in a post-apocalyptic world will require the rediscovery of geology and surveying.

Modern fertilizers provide an optimal balance of these three required nutrients (not unlike the carefully designed diets of top athletes). Using the more rudimentary methods discussed in this chapter, you won't achieve yields as high as the enriched soils of today, but you will be able to preserve the fertility of the land to a good degree during the recovery period.

ONE FEEDING TEN

For a post-apocalyptic society to progress, it absolutely must secure this solid agricultural foundation. If a brutal cataclysm wipes out a great majority of humanity along with the knowledge and skills they hold, the surviving population could be knocked back to a bare subsistence existence, hanging by its fingertips on the cliff edge of extinction. It doesn't matter how much industrial knowledge or scientific inquisitiveness persists through the apocalypse if the survivors are preoccupied with the struggle for mere survival. With no food surplus, there is no opportunity for your society to grow more complex or to progress. And because growing food is so vital, you're much less willing to change what is tried and tested when your life depends on it. This is the food-production trap, and many poor nations today are caught in it. Thus, post-apocalyptic society may stagnate, perhaps for generations, while the efficiency of agriculture is slowly improved until a critical threshold is surpassed when society can begin clawing its way back up to greater complexity.

On the most basic level, a growing population size means more human brains, which can find solutions to problems more quickly. But efficient agriculture offers an even more important opportunity for progress. Once basic food security is assured by efficient means, a civilization can release many of its citizens from toiling in the fields. A productive agricultural system enables one person to feed several others, who are then free to specialize in other crafts and trades.[3] If your brawn is not demanded in the fields, your brain and hands can be put to other uses. A society can economically develop and grow in complexity and capability only once this basic prerequisite has been met—agricultural surplus is the fundamental engine for driving the advancement of civilization. But the benefits of productive agriculture for a rapid reboot of civilization after the apocalypse can be realized only if the excess food can be stored reliably and doesn't rot away uneaten: we'll now turn to food preservation.

---

3 Using many of the advances discussed in this chapter, between the sixteenth and nineteenth centuries the British agricultural revolution achieved a substantially greater production of food while simultaneously becoming less labor-intensive, and the fact that a decreasing proportion of farmers and agricultural laborers was needed to feed everyone else enabled greater urbanization. By 1850, Britain had the lowest proportion of farmers of any country in the world, with only one person in five working the fields to feed the entire nation. By 1880 only one Briton in seven had to work the land, and by 1910 that had fallen to one in eleven. And in developed nations today, which exploit artificial fertilizers, pesticides, and herbicides, as well as enormously labor-efficient technologies like combine harvesters, every agricultural worker grows enough food to feed around fifty others.

CHAPTER 4

FOOD AND CLOTHING

Burg-places broken, the work of giants crumbled.
Ruined are the roofs, tumbled the towers,
Broken the barred gate: frost in the plaster,
Ceilings a-gaping, torn away, fallen,
Eaten by age . . .

UNKNOWN EIGHTH-CENTURY SAXON AUTHOR
LAMENTING ROMAN REMAINS, "The Ruin"

COOKING IS THE ORIGINAL CHEMISTRY in our history—deliberately directing the transformation of the chemical makeup of matter. The crispy browning on the outside of a grilled steak and the golden crust of a loaf of bread are both due to a particular molecular change known as the Maillard reaction. Proteins and sugars in the food react together to create a whole host of new, flavorsome compounds. But cooking serves far more fundamental purposes than simply making food taste more appetizing, and it will form the crux of keeping the survivors healthy and well nourished after the apocalypse.

The heat of cooking kills any contaminating pathogens or parasites, preventing food poisoning from microbes or infection with tapeworm from pork, for example. Cooking also helps to soften tough or fibrous food, and breaks down the structures of complex molecules to release simpler compounds that are more easily digested and absorbed. This increases the nutritional content of much food, allowing our bodies to extract more energy from the same volume of edible matter. And in some cases, such as taro, cassava, and wild potato, prolonged heat inactivates plant poisons, which in the extreme example of cassava would otherwise be lethal from a single meal.

Cooking is only one kind of processing that we apply to food before consumption. The capability to keep food safely for protracted periods beyond its immediate collection is a fundamental prerequisite for the support of civilization. It allows produce to be transported from the fields or slaughterhouses into cities to support dense populations, and enables the stockpiling of reserves for leaner times. Food is spoiled by the action of microbes—bacteria as well as molds—breaking down its structure and changing its chemistry, or releasing waste products distasteful or even outright toxic to humans. The purpose of food preservation is to prevent this microbial spoilage occurring, or at least to delay the process for as long as possible. You pull this off by deliberately modifying the conditions in the food to push them outside the sweet spot for microbial growth. We'll come in a second to a more detailed explanation of how this is actually achieved, but you are essentially trying to exert control over the food's microbiology: preventing any microorganism growth, or even employing some microbes to block other, undesirable strains from gaining a foothold. In some cases, fermentation from microbial growth is encouraged to decompose the complex molecules in food and make nutrients more readily accessible for our own consumption. Biotechnology, therefore, is far from a modern innovation; it is, in fact, one of humanity's oldest inventions.

The development that first endowed us with all of these capabilities—cooking food thoroughly by boiling or frying, fermentation processing, and long-term preservation—was the innovation of firing clay into earthenware pots. This had profound ramifications for us as a species. The human digestive system, unlike the multiple stomachs of ruminants like cows, for example, is unable to break down many food types particularly well, and so we have applied technology to supplement what our bodies can naturally do. Pottery vessels, used as receptacles for food during fermentation or cooking to release further nutrients, therefore serve as additional, external "stomachs"—a technological pre-digestive system.

Modern cuisine—the height of civilized sophistication with all its marinades, confits, and drizzles of reductions—is no more than a superficial adornment upon these fundamental necessities of stopping food from poisoning you and unlocking as much of its nutritional content as possible. This isn't a cookbook, so we won't go into recipes or detailed instructions, but the general principles behind preservation and processing methods are crucial knowledge for a post-apocalyptic recovery.

FOOD PRESERVATION

Preserving food takes into account the environmental conditions that microbes, and indeed all life, need to thrive. But the traditional techniques we'll look at were all developed over long periods by trial and error, long before the discovery of invisible microorganisms causing decay (even the modern practice of canned food was adopted before the demonstration of the germ theory). These techniques were found to work, but without any underlying theory as to why. Retaining this kernel of understanding after the apocalypse (see page 160 for how to build a microscope capable of revealing these microbes) will be enormously beneficial to maintaining a reliable food supply and avoiding infectious disease—both critical to sustaining a population increase after a cataclysm.

Not only does all life on Earth require liquid water to grow and reproduce; organisms can also tolerate only a particular range of physical or chemical conditions. More specifically, the enzymes in a cell— the molecular machinery that drives the reactions of biochemistry and coordinates the processes of life—are active only over particular ranges of temperature, salinity, and pH (how acidic or alkaline a fluid is). Preservation can be achieved by pushing any of these three factors away from the optimum for microbial growth.

The easiest method of preserving food is simply to desiccate it. Without much available water, microbes struggle to grow (this is why it's also critical to dry your harvested grain before storing it in silos). The traditional technique is air- or sun-drying, suitable for fruit such as tomatoes as well as meat to make biltong or beef jerky, but it is a slow process and not suitable for large bulks of food.

Without being commonly considered as desiccated, many other foodstuffs are also preserved by low water availability. Large amounts of dissolved compounds like sugars make a solution very concentrated, which acts to draw water out of microbial cells and stop all but the hardiest strains from growing. This is exactly the principle behind jams: the saccharine fruitiness tastes great on toast in the morning, but the very reason for the creation of preserves in the first place is to protect fruit by the antimicrobial action of the concentrated sugar solution. Sugar can be extracted from tropical sugar cane or the root of the temperate-growing sugar beet by trickling water through the crushed plant to dissolve the sugar and then recovering the crystals of it by drying. Honey is extremely long-lasting for the same reason.

Salt is needed in small amounts for the healthy functioning of the human body—which is why our palate craves it—but a far greater quantity is used for preservation. Salted foods are protected in the same manner as preserves: concentrated briny fluids draw water out of cells and hamper growth. Fresh meat can be effectively preserved by packing it in dry salt for several days, or keeping it submerged in a heavy brine solution—about 180 grams of salt dissolved into every liter of water creates a brine solution roughly five times more concentrated than seawater. Salting has been a crucial preservation technique throughout history, so it is worth looking at in more detail.

In principle, producing salt is childishly simple, provided you're anywhere near the coast. Seawater contains about 3.5 percent dissolved solids, the vast majority of which is common salt (sodium chloride), which can be extracted by evaporating off the water solvent. If you live in sunny climes, you can simply allow seawater to flood shallow pans and evaporate in the heat of the day to leave a crust of salt precipitated behind. In very cold temperatures, you can allow shallow ponds of seawater to freeze, leaving a concentrated brine solution at the bottom. But temperate conditions, as are prevalent across much of Europe or North America over the year, require burning fuel to heat cauldrons of saline to drive off the water. In the case of salt, then, the availability of a valuable commodity is not due to the rarity of the substance itself—three-quarters of the Earth's face is sloshing with saline solution—but to the energetic costs of extracting it in large amounts, or of finding and exploiting minable deposits.[1]

Salting is often used in combination with another preservation technique, whereby naturally toxic antimicrobial compounds are generated and infused into the produce, often meat or fish: the process of smoking. As we'll see in Chapter 5, the incomplete combustion of wood releases a broad suite of compounds, one class of which, creosote, is responsible for the distinctive flavor and decay-inhibiting effect of smoked food. You can jury-rig a small-scale post-apocalyptic smokehouse very easily. Dig a pit for a small fire, with a metal cover, and a shallow trough leading a yard or two to the side, also covered on top with board and then soil, to channel the smoke. At the open end of the covered channel, where the smoke escapes, place a defunct fridge with a hole cut in the bottom. Stock the wire frame shelves with gutted fish, slices of meat, cheese, and so on, and smoke it for several hours.

---

1 Vestiges of the historical importance of salt remain in our language today. Roman soldiers, for instance, were given an allowance to buy salt, which is the derivation of the word "salary."

Acidity is another great ally in resisting the hordes of invading microbes. Vinegar is a weak solution of acetic acid (which we'll come back to later in this chapter) and is very effective as a preservative in pickling. The opposite approach, preserving food with alkalinity, is much less prevalent because it saponifies the fats—see soap-making in Chapter 5—and so grossly changes the flavor and texture of the food.[2]

Rather than adding acid from elsewhere to preserve by pickling, food can also be protected by encouraging the growth of particular bacteria that excrete acidic waste products—allowing food to generate its own preservative. Sauerkraut, Japanese miso, and Korean kimchi are all produced by first using salt to draw out moisture from the vegetables and then allowing fermentation by salt-tolerant bacteria to increase the acidity naturally, transforming the food into an extreme environment and so blocking colonization by other microbes that may cause spoilage or food poisoning.

Yogurt is produced in a similar way, by allowing a culture of lactic acid–releasing bacteria to sour the milk (in general, acids are perceived by the tongue as sour-tasting) in a controllable way. This again creates an internal environment with enhanced acidity that resists colonization by other microbes and so prolongs the consumability of the nutrients by several days. With milk being such a useful source of key nutrients, its preservation is key for survivors of the apocalypse.

Vitamin D is essential for preventing the bone-degradation disease rickets, since it aids the absorption of calcium from food. This vitamin is manufactured by the body when skin is exposed to sunlight, but at

2 One exception is the preparation of corn used traditionally by the native cultures of Mesoamerica. Here the corn is boiled in an alkaline solution, from either slaked lime or ashes thrown into the water, to "nixtamalize" it (from the Nahuatl words for ashes and corn dough). Not only does this improve the flavor, it also makes the crop's vitamin B3 available for absorption by the body. The disease pellagra, caused by deficiency of this vitamin, plagued Europeans and North Americans relying on a staple of corn for two centuries, because they adopted the crop but not the proper technique of preparing the grain for consumption.

northern latitudes, with long, dark winters when people have had to wrap up against the cold, rickets can be a big problem. Few foods naturally contain vitamin D (although oily fish keep the Inuit healthy) so make sure you get enough sunlight and calcium from dairy products: being able to reliably preserve the nutrients in milk will be crucial to healthy habitation of the north.[3]

Butter is a good way of preserving the energy-rich fats of milk by removing much of its water. The essence of butter making is to first extract the fat-rich cream; you can either allow it to naturally rise to the top for a day or so in a cool container, or accelerate the process with a centrifuge (a whirling bucket will do the trick). The process of churning is simply to get the droplets of fat to stick together and exclude the remaining fluid, or buttermilk. This can be achieved by rolling a jar back and forth across the floor, or shaking it, but a more effective post-apocalyptic makeshift solution would be to use an electric drill with a paint-stirring paddle. Strain the butter out of the buttermilk, add salt for preservation, and then knead it until all the water has been squeezed out and the salt mixed throughout.

Yogurt and butter are stable for a few days to around a month, respectively, whereas cheese can safely preserve the nutrients of milk for many months: it is the perfect rickets-busting storage medium. Cheese making is more involved, but the crucial point is to preserve the nutrients in milk by removing its water component. Rennin, an enzyme from the first stomach of a calf, is used to break down the proteins in milk and so curdle it. The curds are strained off and pressed into a solid lump, which is then allowed to mature; it's the action of various fungi that gives different cheeses their characteristic appearance and flavor.

----

3 The landmasses in the northern hemisphere extend much closer to the pole than in the southern hemisphere—Newcastle, Moscow, and Edmonton are far closer to the pole, and so receive less winter light, than anywhere on the southern continents of Africa, Australia, or South America.

PREPARATION OF CEREALS

Let's turn our attention now to the preparation of cereal crops. The prehistoric domestication of wheat, rice, corn, barley, millet, and rye represents one of the crowning glories of human accomplishment. The reproductive strategies of these cultivated strains have been reprogrammed through artificial selection to bear easily recoverable grain—they are the solution that we found to the challenge of consuming grass species without the biological benefit of a ruminant digestion like that of the cows and sheep we husband.

Corn can be cooked and eaten as corn on the cob,[4] and rice can be dehusked and simply boiled or steamed for eating. But the small, hard kernels of most cereal crops cannot be eaten as is, unlike many cultivated fruits or vegetables; they have to be technologically prepared for consumption.

The grain must be pulverized into a fine powder: flour. The simplest method is to place a handful of grain onto a smooth, flat rock on the ground, and then lean forward and use your body weight to crush and grind it beneath a handheld pestle stone. But this is backbreaking and enormously time-consuming labor: a far better system is to mill it between two squat cylindrical stones or steel disks, with the grain introduced (grist for the mill—another common phrase with ancient agricultural roots) into the sandwich through a hole in the middle. The weight of the top millstone provides the crushing pressure, and its rotation works the flour outward to be collected. In this way, the millstone represents a technological extension of our molar teeth, crushing and grinding hard foodstuffs to render them more digestible. You can

---

4 And more than six thousand years ago South American inhabitants discovered how to explosively evert the kernels of particular varieties—now the basis of a billion-dollar cinema-focused market in the US alone.

ease your own manual toil by yoking a draft animal to drive this slow rotation, or even better, harnessing water or wind energy (we'll see how in Chapter 8). Even so, the pulverizing of a harvest's worth of grain will represent an enormous expenditure of energy for a recovering society.

The simplest, but least appetizing, way to consume ground flour is to mix it with a little water into a thick porridge or gruel. But there is a far more tasty, and versatile, starch-delivery mechanism that requires just a little more preparation. Bread is essentially no more than a cooked gruel, but as an effective pathway for nourishment it has underpinned civilization since its very birth. The basic recipe is ludicrously simple: grind some grass seeds into a powdery flour, mix with water into a pasty dough, then roll out and cook slowly, perhaps even just on a hot stone by the fire. This makes an unleavened flatbread, which is still exceedingly common today as chapati, naan, tortilla, khubz, and pita bread.

The type of bread we are most familiar with in the Western world, though, is risen bread, and this requires one further ingredient. Yeast is a microbe, a single-celled fungus not far removed from the toadstools sprouting from a rotting tree trunk, that is applied to the fermentation of flour dough, breathing out carbon dioxide that gets trapped in bubbles to produce a light, fluffy loaf. One particular strain of yeast, *Saccharomyces cerevisiae*, is used for producing almost all leavened bread today. Indeed, you'd do very well to have the presence of mind to rescue a starter stock of this organism, as vital and hardworking in its own way as the ox or horse, before it is lost in the turmoil of the apocalypse; it can be found in supermarkets dried, in packets, but won't persist indefinitely. How might you go about re-isolating bread-making microorganisms from scratch if you have to?

The yeast required for raising bread, as well as other fermentation bacteria, are naturally present on cereal grain and thus also in milled flour. The trick is to isolate these beneficial bugs among all the others that might do you harm: you need to play primitive microbiologist and create a selection process that favors the desired bugs. The guide given here is for isolating the correct microbes for baking a sourdough, the first leavened bread to be baked, around 3,500 years ago in ancient Egypt, and still popular today among craft bakers.

Make up a mixture of one cup of flour (whole-grain is best for this initial process) and half to two-thirds of a cup of water; cover and allow it to sit in a warm place. Check after twelve hours for signs of growth and fermentation, such as bubbles forming. If none are apparent, stir and wait another half day. Once you get fermentation, throw half of the culture away and replace with fresh flour and water in the same proportions, repeating this refill twice a day. This gives the culture more nutrients to reproduce and continually doubles the size of the microbial territory to expand into. After about a week, once you have a healthy-smelling culture reliably growing and frothing after every replenishment, like a microbial pet thriving on the feed left in its bowl, you are ready to extract some of the dough and bake bread.

By running through this iterative process you have essentially created a rudimentary microbiological selection protocol—narrowing down to wild strains that can grow on the starch nutrients in the flour with the fastest cell division rates at a temperature of around 20°–30°C. Your resultant sourdough is not a pure culture of a single isolate, but actually a balanced community of lactobacillus bacteria, able to break down the complex storage molecules of the grain, and yeast living on the byproducts of the lactobacilli and releasing carbon dioxide gas to leaven the bread. Such a mutually supportive marriage between different species is known as a symbiotic relationship and is a common feature of biology from nitrogen-fixing bacteria hosted in the roots of legume plants, to the bacterial digestion assistants in our own gut. The lactobacilli additionally excrete lactic acid (just as in yogurt production), which gives this bread its tasty sour tang, but also act to exclude other microbes from the culture, keeping the symbiotic sourdough community wonderfully stable and resilient against incursion.

Not all flours can be used for leavened bread, however, as it requires the presence of gluten to create a malleable dough able to trap the bubbles of carbon dioxide breathed out by the growing yeast. Wheat grain contains lots of gluten and so makes a divinely light-textured loaf, whereas barley flour has barely any. Barley has a far more pleasing application than daily bread, however.

Yeasts growing in an environment with plenty of oxygen, such as in a dough, are able to break down their food molecules all the way to carbon dioxide (just as human metabolism does). But culture yeasts under anaerobic conditions, with restricted oxygen, and they can only partially decompose sugars, instead releasing ethanol (alcohol) as a waste product: this is the essence of brewing. Since its discovery, alcohol has been helping revelers have a good time, but it has a myriad other uses and is well worth the effort of purifying in the interests of rebuilding civilization. Concentrated ethanol is valuable as a clean-burning fuel (such as in a spirit burner or biofuel car), a preservative, and an antiseptic. It is also a versatile solvent for dissolving a variety of compounds insoluble in water, such as in the extraction of chemicals from plants for perfumery or creating medical tinctures. And when alcohol is exposed to air for a while it turns vinegary, as any wine drinker is surely familiar with after a bottle has been open for a few days. New bacteria colonize the fluid and convert the ethanol into acetic acid: cooking or table vinegar is commonly between 5 and 10 percent acetic acid diluted in water, and more concentrated solutions can be used for pickling, as we have seen.

Unlike the mixed microbial community of a sourdough, the pure yeast culture used in brewing cannot itself break down the complex starch molecules in the grain, which must therefore first be converted into fermentable sugars. The biological function of starch is as an energy source to support the young sprouting plant until it has become established with leaves, and so the grain's own mechanisms are activated to disassemble the starch. The barley grains (or indeed those of any other cereal) are steeped in water and encouraged to germinate for a week in a warm damp room, and thereby break apart their starch into accessible sugars (the starch molecule is a long chain of sugar subunits linked together), before being dried or partially roasted—to vary the color and flavor of the final brew—in a kiln. This malt is then mashed with hot water to dissolve out all of the sugars, and filtered to produce a sweet-tasting wort. The wort is first boiled, both to evaporate off some of the water in order to concentrate the sugars, and also to sterilize it and so offer a blank slate for adding the desired fermentation microbes afterward. Finally the wort is cooled and inoculated with yeast from a previously brewed batch, and then fermented for around a week.

One exceedingly useful item to scavenge from the supermarket shortly after the Fall would be a bottle of craft ale that contains a sediment of live yeast at the bottom, so as to save this handy bug for posterity. But yeasts suitable for brewing are also prevalent in the environment and can be re-isolated using a selection technique similar to that described above. In fact, the pure-culture yeasts used for commercial bread making today are descended from cells originally found in the froths of beer-brewing fermenters, and were isolated using the microbiological tools of agar plates and microscope that are described in Chapter 7. So next time you're warmly tipsy, remember that your brain has been mildly poisoned and impaired by the excrement of a single-celled fungus. Cheers!

Pretty much any sugar source (or starch disassembled back into sugar) can be fermented into an alcoholic product: honey, grapes, grain, apples, and rice are transformed into mead, wine, beer, cider, and sake, respectively. But regardless of the nutrient source, alcohol from fermentation can only reach a concentration of around 12 percent before the yeast cells essentially poison themselves with their own ethanol excretion. The process of purifying alcohol to higher concentrations, by separating ethanol from the water and everything else in the messy ferment, is known as distillation and is another truly ancient technology.

As with extracting salt from saline solution, separating alcohol from the watery soup of the ferment exploits a difference in the properties of the two components—in this case the fact that ethanol has a lower boiling point than water. At its simplest, a still need be no more complicated than that used by Mongolian nomads to make their hooch. A bowl of the fermented mash is held over a fire, with a collection vessel on a ledge above it, and then a third, pointy-bottomed pot full of cold water positioned directly above both; a hood is then draped over the entire arrangement. The fire heats the mash, and the ethanol is driven off first, the vapor condensing on the cool underside of the water pot and running down to drip into the middle dish. Modern laboratories merely replicate this basic setup with dedicated glassware, a thermometer to check that the steam boiling off the mash doesn't exceed 78°C (the boiling point of ethanol), and a gas burner with a controllable air inlet. The efficiency of the process can be improved by using a fractionating column, an upright cylinder packed with glass beads, so that the vapor coming off the mash repeatedly condenses and re-evaporates, further concentrating the alcohol relative to water each time, before a final condenser with a water-cooled jacket collects the distillate.

MAKING USE OF HEAT AND COLD

Finally, we'll look at how the mastery of temperature—using extremes of heat and cold—has become invaluable for food preservation.

The preservation techniques used throughout history—drying, salting, pickling, smoking—are pretty effective but often change the flavor of the food, and are not perfect in maintaining the nutritional content. A new method was devised by a French confectioner in the early years of the nineteenth century: sealing the food in glass jars with a cork stopper and wax, and then standing the jars in hot water for several hours. Soon after, airtight metal cans began to be used (the reason that we use tin cans, or at least tin-coated steel, today is that this is one of the few metals that will not corrode with the acidity of foods).[5] Encouragingly for an accelerated reboot, there was no missing prerequisite technology that prevented the development of canned food centuries earlier in our history—perhaps even skilled Roman glassworkers could have made reliably sealable airtight vessels—so survivors can start canning food soon after the Fall.

The key principle of the canning process is to inactivate microbes already present using heat and apply an airtight seal to prevent any more from recontaminating the food to cause decomposition. A related procedure, called pasteurization, involves briefly heating foodstuffs to 65–70°C so as to deactivate spoilage or pathogenic microbes. This has been particularly effective in treating milk (without curdling it) to prevent the transmission of tuberculosis or gastrointestinal diseases to humans. For the safest preservation, food that isn't already acidic or pickled should be pressure-canned, exposing it to temperatures above the normal boiling point, as this completely sterilizes the contents and kills even temperature-resistant spores of microbes like those responsible for botulism.

That's how high temperatures can be used to preserve vital stockpiles of food for many years. But what about the cold?

As temperature drops, the activity and reproduction of microbes are slowed, as are the chemical reactions that turn butter rancid and

5 The first can openers did not appear until the 1860s, fifty years after the French army began issuing canned food. Soldiers were expected to open their rations with a chisel, or their bayonet, and it was only once cans became widespread among the civilian population that the opener was needed.

soften fresh fruit. The preserving effect of low temperature has been known for a long time. At least 3,000 years ago the Chinese were gathering ice in winter to preserve food in caves through the year, and in the 1800s Norway was a major exporter of its ice to Western Europe. But being able to artificially create cold is a fundamental advance of modern civilization—and is much trickier to pull off than generating heat. The application of the principles known as the gas laws to create refrigerators is handy for keeping fresh food from spoiling rapidly and for freezing for long-term preservation, but can also be applied to the safe storage of hospital stocks of blood or transportation of vaccines, as well as to building air conditioners or distilling air to produce liquid oxygen. We'll take quite a detailed look at how refrigerators work, because it also illustrates an interesting point about the adoption of technology and how a recovering society could end up taking very different paths from our own.

The key operating principle behind refrigeration is that as a liquid vaporizes to gas, it removes the heat required for that transition from its surroundings. This is why our bodies perspire to keep cool, and a low-tech solution for refrigeration is essentially a sweating clay tub. The Zeer pot, common in Africa, consists of a lidded clay tub inside an unglazed larger one, with the gap between them filled with damp sand. As the moisture evaporates it draws heat out of the inner container, lowering its temperature, so the Zeer pot can postpone the spoilage of fruits or vegetables at market by a week or more.

All mechanical refrigerators work on the same basic principle: controlling the vaporization and recondensation of a "refrigerant." Vaporization (boiling) requires heat energy, whereas condensation releases that thermal energy. If you arrange for the vaporization part of the cycle to happen in pipes within an insulated box, you draw heat out of this closed space and so cool the insides, allowing you to export that heat out into the surrounding air through black radiator fins at the back of the appliance.

Practically all modern refrigerators force the condensation step—returning the refrigerant to a liquid so that it can be vaporized again and remove more heat from the compartment—by using an electric compressor pump. But there are alternative methods, the simplest of which is known as an absorption refrigerator (Albert Einstein himself co-invented one version). In this system, a refrigerant such as ammonia is condensed not by pressurizing it but simply by allowing it to dissolve, or be absorbed, into water. The refrigerant is returned to the cycle by heating the ammonia-water mixture to separate the ammonia, which has a far lower boiling point (the same principle of distillation we saw on page 91), using either a gas flame, an electric filament, or just the Sun's warmth. In this way, an absorption refrigerator ingeniously uses heat to keep things cool. Indeed, without the need of an electric motor for the compressor pump, this design has no moving parts, thus slashing the need for maintenance and the risk of breakdown. And it operates silently.

If history is just one damn thing after another, then the history of technology is just one damn invention after another: a succession of gadgets each beating off inferior rivals. Or is it? Reality is rarely that simple, and we must remember that the history of technology is written by the victors: successful innovations give the illusion of a linear sequence of stepping stones, while the losers fade into obscurity and are forgotten. But what determines the success of an invention is not always necessarily superiority of function.

In our history, both compressor and absorption designs for refrigeration were being developed around the same time, but it is the compressor variety that achieved commercial success and now dominates. This is largely due to encouragement by nascent electricity companies keen to ensure growth in demand for their product. Thus the widespread absence of absorber refrigerators today (except for gas-fueled designs for recreation vehicles, where the ability to run without an electrical supply is paramount) is not due to any intrinsic inferiority of the design itself, but far more due to contingencies of social or economic factors. The only products that become available are those the manufacturer believes can be sold at the highest profit margin, and much of that depends on the infrastructure that already happens to be in place. So the reason that the fridge in your kitchen hums—uses an electric compressor rather than a silent absorption design—has less to do with the technological superiority of that mechanism than with quirks of the socioeconomic environment in the early 1900s, when the solution became "locked in." A recovering post-apocalyptic society may well take a different trajectory in its development.

CLOTHING

We've seen how pottery used for cooking and fermentation aids digestion like an external stomach, and the millstone serves as an extension of our molars. In the same way, clothing is another application of technology to enhance the natural biological capability of our own bodies, improving our ability to retain body heat and so spread far beyond the East African savanna.

Until only about seventy years ago—the blink of an eye in the timescale of civilization—we clothed ourselves with natural animal and plant products. The first synthetic fiber, nylon, did not appear until the outbreak of the Second World War. The degree of sophistication in organic chemistry required for re-creating these polymers will remain beyond the grasp of a rebooting society for quite a while. There is therefore a deep link between how we have traditionally fed and clothed ourselves—agriculture involving domesticated plant and animal species provides a reliable food source, but also the fibers that are twisted into cordage or woven into fabrics, and the skin that is turned into leather. And the techniques of spinning and weaving underpin many other fundamental functions of civilization: cord for binding, ropes for construction cranes, canvas for the sails of a ship or the blades of a windmill.

Once the hand-me-down clothes from the past civilization have worn out, the rebooting society will once again need to gather suitable fibers from the natural world. Plant sources include the pithy stems of hemp, jute, and flax (linen); the leaves of sisal, yucca, and agave; and the fluffy fibers surrounding the seeds of cotton or kapok. Animal fibers can be gathered from the hair of pretty much any furry mammal, although sheep or alpaca wool is most common, and one prevalent insect source is the cocoon of the *Bombyx mori* moth: silk. In this way, both a woolly hat and a fine silk dress are composed of protein not much different from steak, whereas a linen jacket or cotton shirt is made of the same basic stuff as newspaper: sugar molecules strung together into cellulose plant fibers.

So what are the basics required for transforming clumps of natural fiber plucked from cotton or shorn from sheep into clothes to keep you alive? We'll start first with the more rudimentary, entry-level techniques, before looking at how these were overhauled by the world-changing mechanization that began with the Industrial Revolution in eighteenth-century Britain. Our focus will be mainly on wool, which in the event of a cataclysm will remain obtainable over a much wider geographical range than alternatives like cotton or silk.

Once the shorn wool has been picked clean of debris and bits of vegetation, it is washed in warm soapy water to remove much of the grease in the fibers. It then needs to be carded: repeatedly combed between two paddles studded with pins, to break up and thin out a tight wad of wool into a soft, fluffy roll of straightened and aligned fibers. This prepared "roving" is now ready for spinning.

The goal of spinning is to turn a fluff of short fibers into a length of strong thread. You can achieve this without any tools at all, by gently tugging at the roving to pull away a loose tangle of fibers and then twisting this between the tips of your finger and thumb into a thin thread. Although it's possible to do this by using your hands alone, it is awfully time-consuming, so ideally you would want to employ some technology to make the task easier. The spinning wheel is able to perform both important functions: drawing out the roving into a thin strand and then spinning it into a sturdy yarn.

The large wheel is operated by hand, or by your foot working a treadle, and is linked by a belt or cord to more rapidly turn the front spindle shaft. The key mechanism here, the spindle flyer, was conceived by Leonardo da Vinci around 1500, and is the rare brainchild of the inventor that was actually constructed during his lifetime. The U-shaped flyer rotates slightly faster than the spindle, and the strands being spun are guided through a row of hooks along one of the arms before slipping off the end and wrapping around the central spindle. This ingeniously simple design is able to simultaneously impart twist to the fibers and wind them into a reel of thread to be used later. Even so, making enough thread with a spinning wheel is so time-consuming that historically it was performed only by young girls or older unmarried women—spinsters.

SPINNING WHEEL, SHOWING THE ROVING BEING FED THROUGH THE ROTATING SPINDLE FLYER ARMS SO THAT IT IS TWISTED INTO THREAD AS IT WINDS ONTO THE SPOOL.

To make a single thread stronger, you can twist it with a second to create a two-ply twine; and, more important, if you twist them in the direction opposite to how they were originally spun, the two intertwined strands will naturally lock together and not unravel. You can repeat this combination process to make ropes thicker than your arm and able to support tons of weight, all from fibers that are very weak individually and no more than a few inches long.

The biggest demand for your spun yarns, however, will be in producing fabrics. Take a close look at the weave of the clothes you're wearing right now. Shirts often have particularly fine weaves, so you'll see the pattern more obviously in a woolen jacket, a T-shirt, or rugged trousers like jeans. You'll also notice a variety of different patterns used in curtains and blankets, bedsheets, duvets, sofa covers, or carpets.

We'll ignore the exact pattern for the time being, but it should be clear that any cloth or fabric is made up of two sets of fibers, at right angles to each other, and interwoven over and under each other. The first set, called the warp fibers, are the main structural components of a fabric and so must be stronger—try two- or four-ply yarns—than the weft fibers that fill in between the parallel lines of the warp and bind them all together.

Weaving is performed on a loom, and the crucial function of any loom is to hold the warp threads in taut parallel lines and then raise or lower different groups of these fibers so that the weft can be threaded between them. The most rudimentary looms consist of no more than two rods—one tied to a tree and the other the ground—holding the warp fibers taut between them, but greater sophistication is offered by a loom with a horizontal frame bearing the warp.

Setting up the loom involves wrapping a continuous twine tightly back and forth straight along its length, creating a grating of neatly parallel warp lines. The crucial component of the loom is the heddle, the device that allows you to separate the warp threads by raising or lowering a subset of them (we'll come back to this in a second). The weft is then passed across the loom through the gap, or shed, that is

WEAVING LOOM. THE HEDDLES ARE LIFTING A SET OF WARP FIBERS TO
ALLOW THE WEFT TO BE PASSED ACROSS THROUGH THE GAP.

created between the parted warps; the set of warp fibers that are raised
is changed and then the weft brought back across again, building up
the mesh of the fabric one row at a time.

Varying the sequence with which subsets of the warp threads are
raised changes the interleaving pattern of the weft and gives you different styles of fabric. In the most basic pattern, the plain weave, the
weft simply passes over and then under a single warp thread each time
to create an even grid of interwoven links—this is the standard weave
for linen. A clever design for a heddle able to achieve this is a long
board with a row of alternating narrow slots and holes, each threaded
with an individual warp fiber. When this rigid heddle is raised or low-

ered only the warp fibers caught in the holes move with it, while those running through a tall slit are unaffected as the heddle slides around them, allowing the weft to be passed under and then over alternating warps.

More complicated weaving patterns demand more complicated heddles than the rigid board. One very versatile system is a series of strings dangling in a row from a horizontal shaft, each with a knotted-loop or metal-eyelet heddle at the same height, so that only those warp fibers threaded through the heddles are lifted when the shaft is raised. Each group of warp fibers is controlled by its own raisable shaft, and the more complex the weaving pattern, the greater the number of separate shafts operating the heddles that is required to adequately control the sequence of warp movements. For example, in a twill weave the weft passes over several warp threads at once (called a float), with the floats staggered between rows to give a diagonal pattern. The reduced number of interlacings, as the warp and weft float over each other, gives a twill weave extra flexibility and comfort, but also allows the yarns to be packed more tightly together, making the fabric more durable. Denim, for example, is a 3/1 twill, whereby the warp and weft threads float for three and then cross over for one.

Whether your garments are stitched from leather or woven fabric, the next problem is how to attach them securely to your body. Disregarding zippers and velcro as too complex to be fabricated by a rebooting civilization, you're low on options for easily reversible fastenings. The best low-tech solution never occurred to any of the ancient or classical civilizations, yet is now so ubiquitous it has become seemingly invisible. Astoundingly, the humble button didn't become common in Europe until the mid-1300s. Indeed, it never was developed by Eastern cultures, and the Japanese were absolutely delighted when they first saw buttons sported by Portuguese traders. Despite the simplicity of its design, the new capability offered by the button is transformative.

With an easily manufactured and readily reversible fastening, clothes do not need to be loose-fitting and formless so they can be pulled over the top of your head. Instead, they can be put on and then buttoned up at the front, and can be designed to be snugly fitted and comfortable: a true revolution in fashion.

Later in the reboot, once the post-apocalyptic population has begun to grow, there will be mounting pressure to begin automating the repetitive and time-consuming processes involved in making fabrics, maximizing the production rate while minimizing the labor required. However, you'll find that both the automation of the different stages— carding, spinning, and weaving—and the application of mechanical power will be much more difficult than for, say, milling grain or pounding wood pulp for paper. Many of the procedures involved in fabric production are exceedingly delicate and suited to the nimble action of fingers, such as spinning a fine thread without snapping it; others, like weaving, demand a complex sequence of actions that must happen at precisely the right moment. All of this is hard to reproduce satisfactorily with rudimentary mechanisms.

The key advance over the basic weaving loom that I described was the invention of the flying shuttle. The simplest way to transfer the weft yarn through the shed gap between raised and lowered warp fibers is by passing a reel of thread from one hand to the other across the loom. But this is a slow action, and also limits the breadth of the fabric to that which can be comfortably reached with your arms. The flying shuttle is a reel of thread encased within a heavy, boat-shaped block that is jerked by a string from side to side of the loom along a smooth runner, playing out weft as it races across. Not only does this innovation allow the weaver to work on a much wider swath of warp fibers, it also greatly accelerates the weaving process and allows the loom to become entirely mechanized, powered by a waterwheel, steam engine, or electric motor, thereby enabling a single weaver to attend to many machines simultaneously. Early power looms could complete a weft row every second, and modern machines convey the weft across the loom more than over 60 miles per hour.

As well as producing food and clothing for yourself, a top priority will be restoring the supply of all the natural and derived substances that are crucial for supporting civilization. Here too the goal is for the post-apocalyptic survivors to learn how to create things for themselves, rather than scavenging from the carcass of our dead society. So let's look now at how to reboot a chemical industry from scratch.

CHAPTER 5

SUBSTANCES

> The shrieks of the birds that nest out there and
> the distant ocean grinding against the ersatz reefs
> of rusted car parts and jumbled bricks and
> assorted rubble sound almost like holiday traffic.
>
> MARGARET ATWOOD, *Oryx and Crake* (2003)

CHEMICALS ARE PRETTY MALIGNED in modern society. We're constantly being told that certain food is healthy because it includes no artificial chemicals, and I've even seen claims for "chemical-free" bottled water. But the fact is that pure water is itself a chemical, as is everything that makes up our bodies. Even before humanity began to settle down and the first cities were founded in Mesopotamia, our lives depended on the deliberate extraction, manipulation, and exploitation of natural chemicals. Over the centuries we've learned new ways to interconvert between different substances, transforming those that can be most easily acquired from our surroundings into those that we need the most, and producing the raw materials with which our civilization has been built. Our success as a species has come not just from mastering farming and animal husbandry or employing tools and mechanical systems to ease labor; it also derives from the proficiency with which we can provide substances and materials with desirable qualities.

The different classes of chemical compounds are like a carpenter's tool kit, each suited to perform a particular task, and we use them to transform raw materials into the products we need, wielding different tools for particular jobs. We'll see that long, chain-like hydrocarbon compounds not only make good stores of energy, but are also water-repellent and therefore crucial for weatherproofing. We'll take a look at different solvents for extraction or purification and investigate how alkalis and their chemical counterparts, acids, have been used throughout history for a number of critical operations. We'll see how some chemicals are able to "reduce" others by stripping away oxygen—a fundamental capability for producing pure metals—while others, known as oxidizing agents, have the opposite action and are able to accelerate combustion, for example. Later in the book we'll examine the chemistry of the processes that make electricity, allow us to capture light for photography, and release a burst of energy for explosives.

I'll concentrate here on some of the most immediately useful substances and processes, a tiny subset of the whole. The fullness of chemistry is a vast network of connections, possible transformations, and conversions between different compounds, and there will be a lot of post-apocalyptic catch-up work in exploring the many features of this territory again, reconnoitering the most efficient methods, rediscovering the ideal ratios to introduce reactants together, and determining the correct chemical formulas and molecular structures.

PROVIDING THERMAL ENERGY

Over time humanity has achieved ever-greater proficiency in controlling and commanding combustion—harnessing fire. Many of the basic functions of civilization rely on chemical or physical transformations driven by heat: smelting, forging, and casting metals; creating glass; refining salt; making soap; burning lime; firing bricks, roofing tiles, and clay water pipes; bleaching textiles; baking bread; brewing beer and distilling spirits; and driving the advanced Solvay and Haber processes we'll return to in Chapter 11. Transient bursts of fire imprisoned inside the pistons of the internal combustion engine power our cars and trucks, and every time you flick on the light switch at home, you're still most likely using fire that has been trapped in a remote location, its energy extracted, changed in form, and then sent down wires to your electric bulb. Our modern, technological civilization is just as dependent on the basic application of fire as our ancestors cooking around the hearth in the earliest human settlements.

Today, much of this required thermal energy is provided either directly or indirectly (via electricity) from the combustion of fossil fuels: oil, coal, or natural gas. Indeed, one of the major enabling technologies for the Industrial Revolution in the eighteenth century was the production of coke from coal and the application of this fuel to many of the processes listed above, in particular to smelting iron and producing steel. Ever since, the progress of our civilization has been powered not by sustainable means, regenerating as much as it consumes, but by plundering these deposits of fossil fuels—energy trapped in transformed vegetative matter accumulated over millions of years.

A society knocked back to basics by an apocalypse may find it difficult to provide for its thermal energy demands once the remnant stocks in gasoline stations and natural-gas storage depots are gone. Much of the easily accessible, high-quality reserves of fossil fuels have already been depleted: the cornucopia of accumulated, ready-to-go energy that gave us an easy ride the first time around is now gone. Oil is no longer found in shallow wells, and coal miners are being forced to delve deeper and deeper into the bowels of the Earth, demanding sophisticated techniques for drainage, ventilation, and support against collapse.[1] Vast reserves of coal do remain globally: the three largest, in the USA, Russia, and China, total more than an estimated 500 million metric tons, but much of the easy-to-get coal has already been got. Some groups of post-apocalyptic survivors may be lucky enough to find themselves near surface coal deposits that can be mined open-pit, but nonetheless civilization restarting after a Fall may be forced into a green reboot.

As we saw in Chapter 1, in the first few decades after the cataclysm, forests will readily reclaim the countryside and even abandoned cities. A small recovering population of survivors will not lack firewood, especially if they maintain coppices of fast-growing trees. The basic principle of coppicing is that once cut down, ash or willow will re-sprout from their own stump and be ready for harvesting again within five to ten years, providing on average five to ten tons of wood every year from one hectare (two and a half acres) of managed forest. Wooden logs are fine for a fireplace warming the home, but for practical applications during the long recovery process you'll need a fuel that burns much hotter than wood. And this necessitates the revival of an ancient practice: charcoal production.

The process is a simple one. Wood is burned with constrained airflow to limit oxygen availability so that it cannot combust completely, but is instead carbonized. The volatiles, such as water and other small, light molecules that turn to gas easily, are driven out of the wood, and then the complex compounds making up the wood are themselves bro-

---

1 Economists rate the quality of a fuel reserve by calculating the Energy Return on Energy Invested (EROEI). This tells you how much usable energy can be gained from a particular deposit relative to all the energy you must expend in mining, refining, and processing it. For example, the first commercially exploited oil fields in Texas in the early 1900s were very easy to harvest and had an EROEI score of around 100—they yielded a hundred times more energy than was consumed in its extraction. Nowadays, as the supplies dwindle, more and more effort must be spent in sucking up (including the hassle of offshore drilling rigs) and processing the remaining drops—the EROEI has dropped to about 10.

ken down by the heat—the wood is pyrolyzed—to leave black lumps of almost pure carbon. Not only does this charcoal burn far hotter than its parent wood—because it's already lost all the moisture, and only carbon fuel remains—but the loss of around half of the original weight also means that it is far more compact and transportable.

The traditional method for this anaerobic transformation of wood—the specialist craft of the collier—was to build a pyre of logs with a central open shaft, and then smother the whole mound with clay or turf. The stack is ignited through a hole in the top, and then the smoldering heap is carefully monitored and tended over several days. You can achieve similar results more easily by digging a large trench and filling it with wood, starting a hearty blaze, and then covering over the trench with scavenged sheets of corrugated iron and heaping on soil to cut off the oxygen. Leave it to smolder out and cool. Charcoal will prove indispensable as a clean-burning fuel for rebooting critical industries such as the production of pottery, bricks, glass, and metal, which we'll come to in the next chapter.

If you do find yourself after the Fall in a region with accessible coalfields, they will again present an irresistible source of thermal energy. A single ton of coal can provide as much heat energy as a year's firewood output of a whole acre of coppiced woodland. The problem with coal is that it doesn't burn as hot as charcoal. It's also pretty dirty—the fumes can taint products made using its heat, such as bread or glass, and sulfur impurities make steel brittle and troublesome to forge.[2] The trick to using coal is to first coke it, echoing the practice of

---

2 In many respects, therefore, charcoal is a fuel superior to coal and is by no means confined to the history books. Brazil, for example, is blessed with abundant timber resources but few coal mines—a situation that is likely to be encountered more broadly in the post-apocalyptic world with regrowth of the forests—and is the largest charcoal producer in the world. Some of it is used in blast furnaces for creating pig iron that is exported to be made into steel for cars and kitchen appliances in the United States and elsewhere. Much of this charcoal is sourced from managed forestry, and so this offers opportunities for "green steel."

turning wood to charcoal. Coal is baked in an oven with restricted oxygen to drive out impurities and volatiles, which, like the products of timber dry distillation, have their own diverse uses and should be condensed and collected.

Combustion also provides light, and while the recovering society restores electrical grids and reinvents the light bulb, survivors will need to rely on oil lamps and candles.[3] Plant oils and animal fats are particularly well suited to serving as condensed energy sources for controllable combustion, due to their chemistry. The main feature of these compounds is their extended hydrocarbon chains: long daisy chains of carbon atoms with hydrogen atoms stuck on at the sides, adorning the flanks like stubby caterpillar legs. Energy is contained within the chemical bonds between the different atoms, and so the long hydrocarbons represent dense reservoirs waiting to be liberated. During combustion, this large compound is ripped apart, and all of the atoms unite with oxygen: the hydrogen atoms combine to form H_2O, water, and the carbon backbone fragments and escapes as carbon dioxide gas. The rapid disassembly of long fatty molecules during oxidation releases a torrent of energy—the warming glow of a candle flame.

An oil lamp can be as basic as a clay bowl with a pinched spout or nozzle, or even just a large shell. A wick, made of plant fiber such as flax or simply rush, draws the liquid fuel up from the reservoir to be evaporated by the warmth of the flame and then combusted. Kerosene has been a common liquid fuel for glass lamps since the 1850s (and today also powers passenger jet planes over the clouds), but it is derived

3 Today, we hold hurricane lamps and candles as reserve technologies, kept as a reliable, easy-to-maintain backup in case the more advanced option fails. But rudimentary technologies also provide a sense of occasion, such as the horse-drawn carriage at a funeral or the candlelit romantic dinner. In this sense, some old technologies never really become obsolete, but persist with a different primary function. For survivors, these methods present hopeful fallback options after the apocalypse.

from the fractional distillation of crude oil and would be difficult to produce after the collapse of modern technological civilization. Any unctuous liquid suffices, though: rapeseed (canola) or olive oil, or even ghee from clarified butter.

A candle can dispense with the container altogether because the fuel itself remains hard until melted in a small pool in the vicinity of the flame—thus a candle is no more than a cylinder of solid fuel with the wick running down the middle. As it burns down, more wick is exposed, producing a larger and smokier flame, unless you periodically snip the wick. The fuss-saving innovation, which didn't occur to anyone until 1825, is to braid the fibers of the wick as a flattened strip, so that it naturally curls over and the excess is consumed by the flame.

Modern candles are composed of wax derived from crude oil, and the availability of beeswax will always be limited, but you can make a perfectly functional candle from rendered animal fat. Boil meat trimmings in salty water, and scoop the hardening layer of floating fat off the surface. Pig lard produces a smelly, smoky candle, but beef tallow or sheep fat are passable. Pour molten tallow into a mold, or even just dip a row of dangling wicks into hot tallow to coat them; allow them to cool and set in the air. Then repeat, building up layer after layer until you have substantial candles.

LIME

The first substance that a recovering post-apocalyptic society will need to begin mining and processing for itself, because of its multitude of functions that are absolutely critical to the fundamental operations of any civilization, is calcium carbonate. This simple compound, and the derivatives easily produced from it, can be used to revive agricultural productivity, maintain hygiene and purify drinking water, smelt metals, and make glass. It also offers a crucial construction material for rebuilding and provides key reagents for rebooting the chemical industry.

Coral and seashells are both very pure sources of calcium carbonate, as is chalk. In fact, chalk is also a biological rock: the white cliffs of Dover are essentially a 100-meter-thick slab of compacted seashells from an ancient seafloor. But the most widespread source of calcium carbonate is limestone. Luckily, limestone is relatively soft and can be broken out of a quarry face without too much trouble, using hammers, chisels, and pickaxes. Alternatively, the scavenged steel axle from a motor vehicle can be forged into a pointed end and used as a drill to repeatedly drop or pound into the rock face to create rows of holes. Ram these with wooden plugs and then keep them wet so that they swell and eventually fissure the rock. But pretty soon you'll want to reinvent explosives and use blasting charges to replace this backbreaking labor.

Calcium carbonate itself is routinely used as "agricultural lime" to condition fields and maximize their crop productivity. It is well worth sprinkling crushed chalk or limestone on acidic soil to push the pH back toward neutral. Acidic soil decreases the availability of the crucial plant nutrients we discussed in Chapter 3, particularly phosphorus, and begins starving your crops. Liming fields helps enhance the effectiveness of any muck or chemical fertilizers you spread.

It is the chemical transformations that limestone undergoes when you heat it, however, that are particularly useful for a great range of civilization's needs. If calcium carbonate is roasted in a sufficiently hot oven—a kiln burning at least at 900°C—the mineral decomposes to calcium oxide, liberating carbon dioxide gas. Calcium oxide is commonly known as burned lime, or quicklime. Quicklime is an extremely caustic substance, and is used in mass graves—which may well be necessary after the apocalypse—to help prevent the spread of diseases and to control odor. Another versatile substance is created by carefully reacting this burned lime with water. The name quicklime comes from the Old English, meaning "animated" or "lively," as burned lime can react so vigorously with water, releasing boiling heat, that it seems to be alive. Chemically speaking, the extremely caustic calcium oxide is tearing the molecules of water in half to make calcium hydroxide, also called hydrated lime or slaked lime.

Hydrated lime is strongly alkaline and caustic, and has plenty of uses. If you want a clean white coating for keeping buildings cool in hot climes, mix slaked lime with chalk to make a whitewash. Slaked lime can also be used to process wastewater, helping bind tiny suspended particles together into sediment, leaving clear water, ready for further treatment. It's also a critical ingredient for construction, as we'll see in the next chapter. It's fair to say that without slaked lime, we simply wouldn't have towns and cities as we recognize them. But first, how do you actually transform rock into quicklime?

Modern lime works use rotating steel drums with oil-fired heating jets to bake quicklime, but in the post-apocalyptic world you'll be limited to more rudimentary methods. If you're really pulling yourself up by your bootstraps, you can roast limestone in the center of a large wood fire in a pit, crush and slake the small batches of lime produced, and use them to make a mortar suitable for building a more effective brick-lined kiln for producing lime more efficiently.

The best low-tech option for burning lime is the mixed-feed shaft kiln: essentially a tall chimney stuffed with alternating layers of fuel and limestone to be calcined. These are often built into the side of a steep hill for both structural support and added insulation. As the charge of limestone settles down through the shaft, it is first preheated and dried by the rising draft of hot air, then calcined in the combustion zone before it cools at the bottom, and the crumbling quicklime can be raked out through access ports. As the fuel burns down to ash and the quicklime spills out the bottom, you can pile in more layers of fuel and limestone at the top to keep the kiln going indefinitely.

A shallow pool of water is needed for slaking the quicklime, and you could use a salvaged bathtub. The trick is to keep adding quicklime and water so that the mixture hovers just below boiling, using the heat released to ensure that the chemical reaction proceeds quickly. The fine particles produced will turn the water milky before gradually settling to the bottom and agglutinating as the mass absorbs more and more water. If you drain off the limewater, you'll be left with a viscous sludge of slaked-lime putty. We'll see in Chapter 11 how limewater is used to produce gunpowder, but let's look here at one particularly useful application of slaked lime: to create a chemical weapon against marauding hordes of microorganisms.

SOAP

Soap can be made easily from basic stuff in the natural world around you and will be an essential substance in the aftermath for averting a resurgence of preventable diseases. Health education studies in the developing world have found that nearly half of all gastrointestinal and respiratory infections can be avoided simply by regularly washing your hands.

Oils and fats are the raw material of all soaps. So, somewhat ironically, if you carelessly splash bacon fat onto your shirt cooking breakfast, the very substance you use to clean it out again can itself be derived from lard. Soap lifts greasy stains from your clothes and washes the bacteria-laden oil off your skin because it is able to mingle comfortably with both fatty compounds and water, which do not themselves mix. It takes a special kind of molecule to display this social-butterfly behavior: one with a long hydrocarbon tail that mixes with fats and oils and a charged head that dissolves well in water. An oil or fat molecule is itself composed of three "fatty acid" hydrocarbon chains all stuck onto a linker block. The key step in making soap, known as the saponification reaction, is to snap the chemical bonds attaching the three fatty acids. A whole category of chemicals known as alkalis are able to do this, "hydrolyzing" the connector bond. Alkalis are the opposites of acids, and when the two meet they neutralize each other to produce water and a salt. Common table salt, sodium chloride, for example, is formed by the neutralization of alkaline sodium hydroxide with hydrochloric acid.

So to make soap, you need to produce a fatty acid salt by hydrolyzing lard with an alkali. While it's true that oil and water don't mix, this fatty acid salt can embed its long hydrocarbon tail into the oil and leave its head poking out to dissolve in the surrounding water. Coated with a fur of these long molecules, a small droplet of oil is stabilized in the midst of the water that rejects it, and so grease can be lifted off skin or fabric and be washed away. The bottle of "invigorating, reviving, hydrating, deep clean sea splash" men's shower gel in my bathroom lists nearly thirty ingredients. But alongside all the foaming agents, stabilizers, preservatives, gelling and thickening agents, perfumes, and colorants, the active ingredient is still a soap-like mild surfactant based on coconut, olive, palm, or castor oil.

The pressing question, therefore, is where to get alkali in a post-apocalyptic world without reagent suppliers. The good news is that survivors can revert to ancient chemical extraction techniques and the most unlikely-seeming source: ash.

The dry residue left behind after a wood fire is mostly composed of incombustible mineral compounds, which give ash its white color. The first step to restarting a rudimentary chemical industry is alluringly simple: toss these ashes into a pot of water. The black, unburned charcoal dust will float on the surface, and many of the wood's minerals, insoluble, will settle as a sediment on the bottom of the pot. But it is the minerals that do dissolve in the water that you want to extract.

Skim off and discard the floating charcoal dust, and pour out the water solution into another vessel, being careful to leave behind the undissolved sediment. Drive off the water in the new vessel by boiling it dry, or if you're in a hot climate, pour the solution into wide shallow pans and allow it to dry in the warmth of the sun. What you'll see left behind is a white crystalline residue that looks almost like salt or sugar, called potash. (In fact, the modern chemical name for the predominant metal element in potash derives its name from this vernacular: potassium.) It's crucial that you attempt to extract potash only from the residue of a wood fire that burned out naturally and wasn't doused with water or left out in the rain. Otherwise, the soluble minerals we are interested in will already have been washed away.

The white crystals left behind are actually a mixture of compounds, but the main one from wood ash is potassium carbonate. If you burn a heap of dried seaweed instead and perform the same extraction process, you can collect soda ash, or sodium carbonate. Along the western shoreline of Scotland and Ireland the gathering and burning of seaweed was a major local industry for centuries. Seaweed also yields iodine, a deep-purplish element that you'll find very useful as a wound disinfectant as well as in the chemistry of photography, which we'll come back to.

If you follow the process described above, you can collect about a gram of potassium carbonate or sodium carbonate from every kilogram of wood or seaweed burned—that is only about 0.1 percent. But potash and soda ash are such useful compounds that it is well worth the effort in extracting and purifying them—and remember that you can use the heat of the fire for other applications first. The reason that timber serves as a ready-packaged stash of these compounds is that over decades of time the tree's root network has been absorbing, from a vast volume of soil, water and dissolved minerals that can then be concentrated with fire.

Both potash and soda ash are alkalis; indeed, the very term derives from the Arabic *al-qalīy*, meaning "the burned ashes." If you now mix your extract into a boiling vat of oil or fat, you can saponify it, creating your own cleansing soap. You can therefore keep the post-apocalyptic world clean and resistant to pestilence with just base substances like lard and ash, and a little chemical know-how.

This hydrolysis reaction is enhanced, however, if you use a more strongly alkaline solution: lye. This is where we return to slaked lime, calcium hydroxide.

You don't want to use slaked lime itself for saponification because calcium soaps form a scum on water rather than a lovely lather. But the calcium hydroxide can be reacted with potash or soda so that the hydroxide swaps partners to produce potassium hydroxide or sodium hydroxide: caustic potash or caustic soda, both of which are traditionally called lye. Caustic soda is powerfully alkaline (it will readily hydrolyze the oils in your skin into human soap, so be extremely careful in handling it) and is therefore ideal for this crucial saponification process, making cakes of hard soap.[4]

Another alkali that is very easy to produce is ammonia. Humans, and indeed all mammals, get rid of excess nitrogen as a water-soluble compound called urea, which we excrete in urine. The growth of certain bacteria converts urea into ammonia—the distinctive stench of which you'll be all too familiar with from poorly cleaned public restrooms—and so the crucial alkali ammonia can also be produced by distinctly low-tech means: fermenting pots of piss. This was historically a crucial process for the production of clothes dyed blue with indigo (traditionally the blue of jeans). We'll return to the diverse uses of ammonia later.

Saponification of fat molecules will also give you another useful byproduct. The chemical component of the lipid that acts as the linker block grasping the three fatty acid tails, glycerol, is left behind after lard is transformed into soap. Glycerol is itself fabulously handy, and

---

4 WARNING: Never use aluminum pots or utensils for creating soap. Aluminum reacts vigorously with strong alkalis to release explosive hydrogen gas.

can be easily extracted out of a lathery soap solution. The fatty acid salts of the soap itself are less soluble in brine than in fresh water, and so adding salt will cause them to sediment out as solid particles, leaving behind the glycerol in the fluid. Glycerol is a key raw material for making plastics—and explosives (which we'll come to in Chapter 11).

The hydrolysis reaction that transforms animal fats into soap is also used for making glue. It's created by boiling skin, sinew, horns, or hooves: anything that contains tough connective tissue made of collagen, which disintegrates to gelatin. This dissolves in water, so can be formed into a gloopy, tacky paste that then dries hard and firm. The necessary hydrolytic breakdown of collagen is much faster under strongly alkaline or acidic conditions—another application for lye, or for acids (which we'll come to in a bit).

WOOD PYROLYSIS

Wood can offer so much more than just carbon fuel and alkali from its ashes. In fact, wood was once the major source of organic compounds—providing chemical feedstocks and precursor substances for a vast array of different processes and activities—and was superseded only in the late nineteenth century by coal tar and the subsequent development of petrochemicals from crude oil. In a post-apocalyptic world, therefore, where you may well find yourself without accessible coal or a continuing supply of oil, these older techniques will support the rebooting of a chemical industry.

The whole point of charcoal making is to drive out the volatiles from wood to leave a hot-burning fuel of almost pure carbon, but these volatile "waste" products are in fact very useful. And with a little refinement of charcoal production, these escaping vapors can also be captured. By the second half of the seventeenth century, chemists had noticed that burning wood in a closed container released flammable

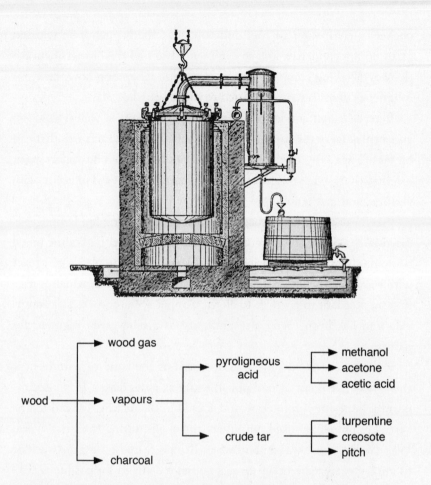

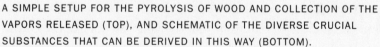

A SIMPLE SETUP FOR THE PYROLYSIS OF WOOD AND COLLECTION OF THE
VAPORS RELEASED (TOP), AND SCHEMATIC OF THE DIVERSE CRUCIAL
SUBSTANCES THAT CAN BE DERIVED IN THIS WAY (BOTTOM).

gas and also vapors that could be condensed back into a watery fluid.
These products came to be known as pyroligneous (a Greek-Latin hybrid word for fire and wood) and are a complex mixture of many different compounds. An ideal stepping-stone for a recovering society to
leapfrog to would be to bake wood in a sealed metal compartment,
with a side pipe drawing off the released fumes and coiling through a

bucket of cold water to cool and condense the vapors. The released gases do not condense, and so can be used to fuel the burners beneath the wood-baking compartment. We'll see in Chapter 9 how these pyroligneous gases can even be used to fuel a vehicle.

The collected condensate readily separates into a watery solution and a thick tarry residue, both of which are complex mixtures that can be teased apart by distillation, as described earlier. The watery part, originally termed pyroligneous acid, is mainly composed of acetic acid, acetone, and methanol.

Acetic acid, as we've seen, can be used for pickling food: vinegar is essentially a dilute solution of acetic acid. It reacts with alkaline metal compounds to produce various useful salts. For example, it can react with soda ash or caustic soda to produce sodium acetate, which is useful as a mordant to fix dyes to cloth. Copper acetate works as a fungicide and has been used since antiquity as a blue-green pigment for paints.

Acetone is a good solvent and is used as the base for paints—it is the distinctive scent of nail polish—and as a degreaser. It is also important in plastics production and is used in the making of cordite, the explosive propellant used for bullets and shells during the First World War. In fact, there was a point when Britain feared losing the war due to an acute shortage of acetone. The huge demand for cordite far exceeded what could be produced by the dry distillation of wood, even with imports of the solvent from timber-rich countries like the United States. Production was maintained by the invention of a new technique, using a particular bacterium to secrete acetone during fermentation, and huge amounts of horse chestnuts collected by schoolchildren as the feed.

Methanol, originally known as wood spirits, is produced in large amounts by the dry distillation of wood: every ton of timber yields about ten liters. Methanol is the simplest alcohol molecule: it contains only one carbon atom, whereas ethanol, or drinking alcohol, is built around a backbone of two. Methanol can be used as a fuel and a solvent; it functions as antifreeze and is also crucial in the synthesis of biodiesel, which we will come to in Chapter 9.

The crude tar sweated out of the roasted wood can also be separated by distillation into its major constituents: thin, fluid turpentine (floats on water); thick, dense creosote (sinks in water); and dark, viscous pitch. Turpentine is an important solvent, used historically for pigments, and we'll come back to this in Chapter 10. Creosote is a fantastic preservative, and when painted on or soaked into wood protects it against the elements and rot. It also acts as an antiseptic, inhibiting microbial growth and preserving meats: it is responsible for the distinctive flavor of smoked meats and fish. Pitch is the gloopiest of the extracts, a viscous mixture of long-chain molecules, and its flammability is ideal for soaking into wooden rods to make torches. This tarry substance is also water-repellent and useful for sealing buckets or barrels; it has been used for millennia to caulk the seams between the wooden slats of a boat's hull.

The timber of any tree will provide differing quantities of these crucial chemicals by dry distillation, but resinous hardwoods, including conifers such as pines, spruces, and firs, yield more pitch. Birch bark is a particularly good source of pitch that has been used since the Stone Age to stick fletching feathers to arrows. Indeed, if it is only the pitch you're after, you can collect it as it oozes out of resinous wood baked in a kiln, or even just in a tin box tossed on a fire.

Distillation is such a universally useful technique for separating a blend of fluids, exploiting the principle that different liquids boil at specific temperatures, that a recovering society would do well to master it as early as possible. Distillation fractionates, or separates, the various products of heat-decomposed wood and extracts concentrated alcohol from a fermented slop, as we've already seen; it also teases apart crude oil into a diverse selection of different constituents, from thick viscous asphalt to light volatile components like gasoline. And once a certain level of industrial capability is achieved, even air itself can be distilled. The gas mixture is chilled to around -200°C by using a repeated expansion and cooling process and is held in a vacuum-insulated capsule, like a giant thermos flask for taking coffee on a hike. The liquid air is then allowed to warm, and as each separate gas boils off it is collected, the pure oxygen used, for example, for hospital breathing masks.

ACIDS

So far we've focused mainly on alkalis, as strong varieties are relatively easy to make. Acids, their chemical counterparts, are just as common in nature, but particularly potent kinds are harder to come across than lyes and have been significantly exploited only more recently in history. We have seen how a variety of plant products can be fermented to produce alcohol, and that this ethanol can in turn be oxidized by exposure to air to produce vinegar. Acetic acid was the earliest acid available to humanity, and for the great majority of history it was also our only option. Civilization has been able to choose from a selection of alkalis—potash, soda ash, slaked lime, ammonia—but for millennia our chemical prowess was limited by wide availability of but a single weak acid.

The next acid to be exploited by humanity was sulfuric acid. This was initially baked out of a rare glass-like mineral called vitriol, and later mass-produced by burning pure yellow sulfur with saltpeter (potassium nitrate) in steam-filled, lead-lined boxes. Today we make sulfuric acid as an offshoot of scrubbing oil and natural gas to remove the sulfur contaminants. So in a post-apocalyptic world you might be caught in the middle: unable to create this crucial, potent acid using traditional methods, as elemental sulfur has long since been removed from surface volcanic deposits, and incapable of pulling off more advanced techniques without the specific catalyst needed.

The trick is to employ a chemical pathway that was never used industrially in our development. Sulfur dioxide gas can be baked out of common pyrite rocks (iron pyrite is notorious as fool's gold, and pyrites also form common ores of lead and tin) and reacted with chlorine gas, which you get from the electrolysis of brine (see page 232), using activated carbon (a highly porous form of charcoal) as a catalyst. The resulting product is a liquid called sulfuryl chloride that can be concentrated by distillation. This compound decomposes in water to form sulfuric acid and hydrogen chloride gas, which should itself be collected and dissolved in more water for hydrochloric acid. Luckily, there is also a simple chemical test for whether a rock is a pyrite mineral (a metal sulfide compound): dribble a little dilute acid on the rock, and if it fizzes and gives off the stench of rotting eggs, you've got what you're after (but hydrogen sulfide gas is poisonous, so don't sniff too much!).

Today, more sulfuric acid is manufactured than any other compound—it is the linchpin of the modern chemical industry, and will also be crucial in accelerating a reboot. Sulfuric acid is so important because it's good at performing several different chemical functions. Not only is it potently acidic, it is also strongly dehydrating and a powerful oxidizing agent. Most of the acid synthesized today is used to produce artificial fertilizers: it dissolves phosphate rocks (or bones) to liberate the crucial plant nutrient phosphorus. But its uses are virtually limitless: preparing iron gall ink, bleaching cotton and linen, making detergents, cleaning and preparing the surface of iron and steel for further fabrication, creating lubricants and synthetic fibers, and serving as battery acid.

Once you've reacquired sulfuric acid, it serves as a gateway to the production of other acids. Hydrochloric acid is produced by reacting sulfuric acid with common table salt (sodium chloride), and nitric acid comes out of the reaction with saltpeter. Nitric acid is particularly useful because it is also a very potent oxidizing agent: it can oxidize things that sulfuric acid can't. This makes nitric acid invaluable for creating explosives as well as for preparing silver compounds for photography—two key processes that we will return to later.

CHAPTER 6

MATERIALS

There was on this continent a more advanced
civilization than we have now—that can't be
denied. You can look at the rubble and the rotted
metal and know it. You can dig under a strip of
blown sand and find their broken roadways. But
where is there evidence of the kind of machines
your historians tell us they had in those days?
Where are the remains of self-moving carts, or
flying machines?

WALTER M. MILLER JR, *A Canticle for Leibowitz* (1960)

AS IS OBVIOUS FROM THE LAST CHAPTER, it is difficult to overstate the sheer usefulness of wood. Its chemical potential aside, timber is one of the most ancient building materials, providing beams, planks, and poles for construction. The particular qualities of different trees are suited to different applications, and there is an enormous amount of accumulated knowledge that would need to be rediscovered by a fledgling civilization after the apocalypse. For example, elm wood's tough, interlocked fibers resist splitting, and it's therefore ideal for cart wheels. Hickory is particularly hard and thus suitable for the gear teeth of the power mechanisms in windmills and water mills. Pine and fir trees grow exceptionally straight and tall, and so make perfect ships' masts.

Beyond these mechanical properties, wood fires will keep the cold at bay once central heating systems have died, and will cook your food to inactivate microbial contamination and help release nutrients. The last chapter showed how to collect the vapors anaerobically baked out of timber to yield a selection of crucial substances: feedstock for rebooting a chemical industry. We've also seen how the resultant charcoal is ideal for filtering drinking water once the taps have run dry and bottled water has disappeared from supermarket shelves. Wood also provides hot-burning fuel for kilns firing pottery and bricks, for making glass, and for smelting iron and steel.

Immediately after the apocalypse you'll be able to simply occupy existing buildings, repairing and patching them up as best you can. But all uninhabited and untended buildings will inexorably decay and collapse over the first decades, and as the surviving population grows and needs new homes, you'll probably find it much easier to construct anew than try to restore the rotting shells of the old civilization. And to do that, you'll need to learn the basics. Brick, glass, concrete, and steel are the literal building blocks of our civilization. But they all come from the humblest of beginnings: mucky earth, soft limestone, sand, and rocky ore that we dig from the ground and transmute with fire into the most useful materials of history. We can see this process most easily with clay, which is shaped and formed while soft and malleable before being heated in a kiln into a hard ceramic. We deliberately change the properties of a substance to suit our application.

CLAY

It is easy to overlook clay in our modern lives—it's perhaps something you associate only with art lessons at school. But the truth is that pottery played an utterly pivotal role in creating the prerequisite conditions for the founding of civilization itself. Lidded receptacles fashioned out of clay enable food to be stored; protect it from pests and vermin; allow for cooking, preservation, and fermentation; and make foodstuffs far more portable for both traveling and trade. Clay formed into blocks and then fired to make bricks also provides a fabulous building material: the fabric of towns, mills, and factories.

Clay beds are exceedingly widespread, and lie beneath the topsoil in many areas of the world. Clay is made up of very fine particles of aluminosilicate mineral—sheets of aluminum and silicon, each bound to oxygen—weathered out of rocks and often transported over great distances by rivers or glaciers before being deposited. Various kinds of clay can therefore be simply dug out of pits in the ground and formed by hand. The most rudimentary receptacle can be formed from a moist ball of clay by pinching into the middle with your thumbs and smoothing into a round bowl. But for far more control over the process, you'll want to redevelop the potter's wheel. The earliest kind was simply a freely rotating disk, so that the workman could turn the piece around as he worked. The "modern" potter's wheel, at least 500 years old and perhaps much more ancient, uses a spinning flywheel, such as a heavy rounded stone, to store rotational momentum and keep the piece turning smoothly as the potter works on it. The wheel is spun up every now and then with a push or kick, or, if you can scavenge one after the Fall, an electric motor.

Dried clay is relatively durable, but ideally you want to fire it to create ceramic. At temperatures between about 300° and 800°C, the water is driven irreversibly out of the clay structure, and the mineral plates lock together but remain porous. Heat it even further, to above 900°C, and the clay particles themselves begin to fuse together, and minor impurities in the clay melt. These vitrifying compounds soak throughout the piece, and when cool they solidify into a glassy matrix, firmly fusing together the clay crystals and filling any gaps to form a hard and watertight material. Deliberately dunking the piece in such substances before the high-temperature firing to seal the surfaces is the art of glazing. You can even just toss some salt into the kiln: the withering heat dissociates the compound, and sodium vapor mingles with the silicon in the clay to form a glassy coating (although noxious chlorine gas is released in the process). This method was historically employed as an easy way to waterproof clay pipes to be used for water distribution or sewer systems.

Fired clay is not just hard and watertight, it is also exceedingly heat-resistant. The aluminosilicate has an extremely high melting point, and since the constituents are already bonded to oxygen, the mineral does not combust when hot. Firebricks are therefore the perfect material to line kilns and furnaces. In order to contain fire, and therefore be able to technologically employ it, you need a substance that can insulate the heat inside but is also able to resist the temperature itself. This is a great example of a recovering civilization pulling itself up by its own bootstraps: baking clay in a large fire into a refractory material enables survivors to build further kilns to fire yet more bricks. The story of civilization itself has been an epic of the containment and harnessing of fire with ever greater finesse to attain ever higher temperatures: from the cooking campfire to the pottery kiln, the Bronze Age smelter, the Iron Age furnace, and the blast furnace of the Industrial Revolution—and it is refractory bricks that have enabled all of this.

Fired clay is also used very commonly as a structural material. In drier climates you can get away with building a rudimentary wall out of sun-dried mud—adobe—but this is at risk of being washed away in a heavy downpour. A far more resilient brick is made by taking a few generous handfuls of clay, squashing it into a cuboidal shape in a mold, and then baking it in a kiln to drive the chemical transformations for a hard, durable ceramic. But you're going to need more than handfuls of clay to rebuild civilization. For a sturdy wall, the rows of bricks will need to be glued together—and for that, we come back to lime.

We saw in Chapter 5 that the first material you're likely to need to start mining again, once the remnant commodities left behind by our current society have been depleted, is limestone. We know limestone plays a central role in synthesizing many of the crucial substances needed by a civilization. Now we'll take a look at how the same wonder material will form the basis of rebuilding in the aftermath. Limestone blocks are useful as a construction material—as is its metamorphic product marble, formed from limestone pressure-cooked deep underground—but it is what this rock can be turned into that is so useful for rebuilding.

Slaked lime is able to transform from a spreadable paste back into a material set hard as stone. Mixed with a little sand and water, slaked lime forms mortar, which has been used to firmly stick bricks together into sturdy load-bearing walls for thousands of years. Mix it with less sand, and perhaps stir in some fibrous material like horsehair, and you have a plaster for spreading as a smooth finish on walls.

Lime mortars have been used for millennia, but it was a substance first mass-produced by the Romans that changed the nature of building. The Romans noticed that *cementum* made by mixing slaked lime with volcanic ash, known as pozzolana, or even pulverized brick or pottery, sets far more quickly than lime mortar and is several times stronger. And with the fabulously strong mineral glue that is cement, you can do far more than just stick together ordered rows of bricks. You can also bond jumbled aggregations of rocks or rubble—that is, you can make concrete. This revolution in construction technology allowed the Romans to build awe-inspiring structures such as the Collosseum and the vast bulging roof of the Pantheon in Rome, which is still the largest single-piece concrete dome in the world.

But it is another, almost magical, property of cement that really helped build the trading and naval prowess of the Roman empire: concrete made with pozzolana or crushed earthenware sets even when completely submerged in water. Unlike lime mortar, cement is said to be hydraulic and sets along a different chemical route. The volcanic ash contains alumina and silica—already discussed above as constituents of clay—which chemically react with the slaked lime to form an exceedingly strong material as they hydrate.

Hydraulic materials led to an important technological advance. Pozzolanic cement spurred a revolution in Roman marine construction, for rather than simply sinking large stone blocks into the water, the Romans could now pour concrete for freestanding structures directly into the sea to create quays, breakwaters, seawalls, and lighthouse foundations. This technology enabled them to build ports wherever needed for military or economic reasons, including in regions with very few natural harbors, such as the north coast of Africa. Thus, Roman ships came to dominate the Mediterranean.

This crucial knowledge about strong cements, versatile concrete, and watertight plasters was nearly lost to history with the fall of the Roman Empire. No medieval sources mention cement, and the great Gothic cathedrals were built using only lime mortar. However, knowledge seems to have been preserved in some places, as hydraulic cement was used in a number of fortresses and harbors constructed throughout the Middle Ages.

But it was in 1794 that the modern method for producing cement was invented. "Ordinary Portland cement" does not exploit volcanic heat like the Roman pozzolana variety, but bakes a mixture of limestone and clay in a specialized kiln at around 1,450°C. The hard clinker produced is then ground up with a small amount of the soft, pale mineral gypsum—also used for plaster of Paris, and in setting broken limbs in a plaster cast—which helps to slow the curing process and gives you more working time with the wet cement.

Now, I know that concrete is a horrifyingly dull and gray building material, and that there have been some architectural abominations constructed with it. But let's step back and consider for a second what truly remarkable stuff it really is. Concrete is essentially man-made rock. And the recipe is beguilingly simple: stir together one bucket of Portland cement with six buckets of sand or gravel, and enough water to make a thick gloop. Pour this liquid stone into a wooden shuttering constructed to make any shape you fancy, and then wait for it to set into an incredibly hard and durable material. It's not difficult to see why concrete allowed rapid urban regeneration following the devastation of the Second World War and is still the prime ingredient for city-building today—an icon of the modern age, even though the basic process was invented more than two millennia ago.

The problem with concrete, however, is that while it's incredibly strong when compressed in foundations or columns, it's very weak when under tension. It catastrophically cracks when forces act to stretch it, which stops it from being used for large structural elements like beams, bridges, or floors of multistory buildings. The solution is to embed steel rods in the concrete. Their individual material properties perfectly complement each other: the compressive strength of concrete combines with the tensile strength of steel. This reinforced concrete was hit upon in 1853 by a plasterer who inserted straightened metal barrel hoops into concrete floor slabs as they set. And it is this final innovation that really unlocks the potential of concrete for aiding reconstruction after the apocalypse.

Concrete is a wonderfully versatile construction material, but it is ceramic bricks, with their refractory properties, that you will need to use to contain high temperatures and so achieve the skills of metallurgy.

METALS

Metals offer a suite of properties not provided by any other materials. Some are exceptionally hard and strong, making them ideal for tools, weapons, or structural components like nails or whole girders. But, unlike brittle ceramics, they also exhibit plasticity—if stressed they deform rather than shatter, and they can be pulled into thin wire suitable for fastening, fencing, or conducting electricity. Many metals can also resist very high temperatures, making them ideal for high-performance machinery.

What you ideally want to be able to reattain as rapidly as possible after the Fall is the mastery not just of iron, but of its carbon alloy, steel. Steel contains a blend of iron and carbon atoms and is so much more than the sum of its parts. The inclusion of carbon changes the metal's properties substantially, and by varying the proportion of carbon soaked into the alloy recipe, you can control the strength and hardness of the steel to suit different applications.

We'll see later how to create iron and steel from scratch, because in the immediate aftermath you'll be able to easily scavenge them. Salvaged items can be repurposed by relearning the traditional skills of the blacksmith: working by an open hearth, or forge, to keep the workpiece glowing hot as you reshape it between hammer and anvil. The reason that we've been able to exploit hard iron through the history of civilization is that it temporarily changes its physical properties when hot, softening to become malleable enough to be beaten into shape, rolled into sheets, or drawn into pipes and wires. This is a fundamental point, because it means you can use iron tools to work on iron to produce more tools.

The crucial knowledge for fully exploiting iron for tools is the principle of hardening steel—quenching and tempering. Steel is hardened by heating it red-hot so that the internal structure of iron-carbon crystals converts to a particularly rigid conformation (which is nonmagnetic, allowing a simple test during heating). If allowed to cool slowly afterward, though, this crystal reverts back to its previous form, and so you need to chill it rapidly to essentially freeze the desired structure: quenching the piece by dunking it, sizzling, into water or oil. However, a hard substance is also brittle—and a steel hammer, sword, or spring that shatters is useless—so after quenching a piece needs to be tempered. You do this by reheating to a lower temperature for a period of time so that a proportion of the crystal structure relaxes: you are deliberately trading off some of the strength to return a little flexibility to the material. Tempering allows you to tune the material properties of steel, a crucial ability for designing the appropriate metal for the intended task.

Another key technology, only more recently available, is welding: gluing metal together with molten metal. Acetylene yields the hottest flame of any fuel gas, reaching over 3,200°C when it is burned in a stream of oxygen. A welding torch can be created by separately controlling the pressurized flow of oxygen and acetylene gas through a lit nozzle. Pure oxygen can be produced by electrolysis of water (Chapter 11), or later in the reboot by distilling liquefied air. Acetylene is released by the reaction of water with lumps of calcium carbide, which is itself made by heating quicklime and charcoal (or coke) together in a furnace: substances that we have already covered. As well as being useful for metal joining, an oxyacetylene flame can also be wielded as a cutting torch for steel, a jet of oxygen combusting the hot metal out in a neat line.

An even higher temperature of around 6,000°C is generated by an electric arc welder: brandishing the power of lightning. Rigging up a row of batteries, or a generator, will produce enough of a voltage that a constant spark, or arc, will leap between the target metal and a carbon electrode, to weld or cut as the electrode is drawn across the surface. Such jury-rigged oxyacetylene torches or arc cutters would be indispensable equipment for salvage crews sent into the dead cities to disassemble the ruins and scavenge the most valuable materials. One very effective way for melting down scrap steel for recycling is the electric arc furnace. It's essentially a giant arc welder: large carbon electrodes surge electricity through the metal to melt it, with limestone flux used to remove impurities as slag on top and the molten steel being poured as if from a kettle. Running an arc furnace from renewable electricity would be an important technology to try to leapfrog to in order to relieve demands on fuels for thermal energy in the post-apocalyptic world.

But retaining access to metals as a class of materials is only half of the story: you'll also need to be able to adeptly work them into the forms you need. If you can't salvage any working versions of the essential machine tools for this, how might you construct them from scratch?

An incredibly elegant example has been provided by a machinist in the 1980s who created a fully equipped metalworking shop—complete with lathe, metal shaper, drill press, and milling machine—starting with little more than clay, sand, charcoal, and some lumps of scrap metal. Aluminum is a good choice, as it has a low melting point for easy casting and is very corrosion-resistant, so it will be scavengeable even long after the apocalypse.

The heart of this phenomenal project is a small-scale foundry, consisting of a salvaged metal bucket fitted with a refractory inner lining of clay, and fired by charcoal intensified with a stream of air through the bucket side. This backyard furnace is more than sufficient to melt the scavenged aluminum, and the molten metal can then be poured to cast a whole variety of machine components. Casting molds can be constructed from fine sand mixed with clay as a binder and a little water, packed around carved patterns in a two-part wooden frame.

The first machine to create is a lathe. A simple lathe is composed of a long, flat beam called the bed, with a headstock fixed at one end and a tailstock at the other that can be unlocked to slide left and right

RUDIMENTARY FOUNDRY: MELTING SALVAGED ALUMINUM IN A SMALL-SCALE
FURNACE (TOP) TO BE CAST IN A SAND MOLD (BOTTOM).

along the bed track. The workpiece is attached to the spindle on the
headstock—perhaps by bolting to a faceplate, or gripped in a chuck

LATHE WITH THE HEADSTOCK AND ROTATING SPINDLE TO HOLD THE
WORKPIECE ON THE LEFT; TAILSTOCK ON THE RIGHT; AND, IN BETWEEN,
THE MOVABLE CARRIAGE BEARING THE CUTTING TOOL.

with movable jaws—and then the whole piece spun around this center, driven by a pulley or gear system from whatever motive force you've harnessed (waterwheel, steam engine, or electric motor). The tailstock can be used to support the other end of the workpiece, sliding along the bed to accommodate different lengths, or else to bear a tool like a drill to bore through the center of the workpiece as it turns. A carriage also slides along the bed, mounted with a cutting tool on a cross slide so that it can be precisely positioned around the workpiece, shaving into it as it turns to create any desired profile. Astoundingly, not only is the lathe capable of duplicating all of its own components to create more lathes, but starting from absolute scratch, you can even produce, during the rudimentary stages of construction of your first lathe, the remaining components needed to complete it.

In order to cut precisely spiraling threads on your workpiece, you would want to fit a long lead screw alongside the bed to smoothly move the carriage, and ideally couple this with gears to the headstock spindle to perfectly coordinate their motion. In a post-apocalyptic world you would really hope to be able to scavenge a ready-made long, threaded screw, as cutting a thread with a constant pitch is fiendishly difficult otherwise. In our history it took a long process of reiterative refinement to make the first precise metal screw thread, from which many others can then be constructed, and you would want to avoid having to repeat this.

Once you have a lathe you can use it to construct the parts of other, far more complex machine tools, such as the milling machine. Whereas the lathe applies a tool to a rotating workpiece, the milling machine bears a rotating tool against the workpiece, and is exceedingly versatile—once you have a milling machine you can create pretty much anything else. So this demonstration is a microcosm of the history of technology itself: simple tools making more complex tools, including more precise versions of themselves, and repeating this cycle to ratchet upward.

But what if you can't find any ready-purified metals for forging or casting, or you've already used all that was scavengeable? How do you get metal out of rocks in the first place? The general principle of smelting is to remove the oxygen, sulfur, and other elements the metal is compounded with in the ore. This requires a fuel to attain high temperatures, a reducing agent, and a flux. Charcoal (or coke) performs admirably for the first two functions: it burns fiercely, and as it combusts in the smelter it releases carbon monoxide, a powerful reducing agent that strips the oxygen away to leave pure metal. The overall blueprint for a rudimentary iron-smelting furnace is similar to that of a kiln for burning lime. The furnace is charged with layers of charcoal fuel and the crumbled iron ore rock. Some limestone mixed in with the ore serves as a flux, lowering the melting point of the refractory gangue (the worthless rocky stuff) so that it turns fluid in the furnace, absorbing the impurities away from the metal. The flux forms a slag that is drained away, and your metallic prize can be extracted from the furnace.

If your furnace doesn't operate hot enough to melt the iron formed, you'll need to retrieve the solid metal as a spongy lump and then batter and pummel it on an anvil to fuse the iron together and squeeze out the remaining slag. To become hard enough to be useful for tools, this pure, wrought iron must be fiercely heated once more with charcoal to make it absorb some carbon—to form steel—and then worked on the anvil again. Repeatedly folding and reflattening it, you'll essentially stir the solid material to create a uniform steel that can then at last be forged into its final form. This is backbreaking work for the smith, and the rate of steel production is severely limited. The key to modern civilization is efficiently making steel in bulk. Here's how you do it.

The solution is to force a powerful stream of air up the furnace stack to ferociously intensify combustion. The Chinese invented the blast furnace by the fifth century BC (more than 1,500 years before it appeared in Europe), and later improved the design using waterwheel-driven piston bellows. For even greater efficiency at achieving high temperatures, preheat the ingoing air blast using the hot, combustible waste gases escaping out of the furnace flue. The freshly smelted iron in the blast furnace absorbs lots of carbon, which acts to lower its melting point to about 1,200°C. The metal liquefies completely and can be run out of the bottom of the furnace, along channels cut in the floor, to cool in a row of ingot molds. The result is pig iron—so named because medieval foundry workers thought the molds looked like newborn piglets suckling on a sow.

This high-carbon iron, with its lowered melting point, can be re-melted and poured into a mold like hot wax. Cast iron is therefore greatly convenient for quickly fabricating items like cooking pots, pipes, and machinery parts, and the Victorians mass-produced cast-iron girders. But cast iron also has one great disadvantage: the high carbon content makes the metal brittle, and cast-iron bridges, for example, have the nasty habit of collapsing if their structural components are bent or stretched.

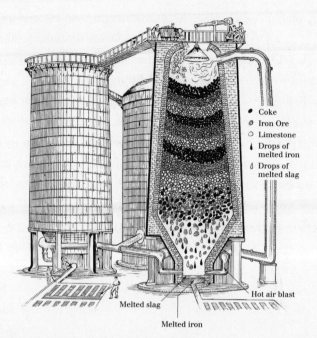

Coke
Iron Ore
Limestone
Drops of
melted iron
Drops of
melted slag

Hot air blast

Melted slag

Melted iron

BLAST FURNACE FOR SMELTING IRON. THE ORE, FUEL, AND FLUX ARE FED IN
AT THE TOP, AND AN INTENSE STREAM OF HOT AIR IS FORCED UP THE STACK
FROM THE BOTTOM.

The innovation that really made possible the later stages of the Industrial Revolution was a means for easily transforming blast-furnace pig iron into steel. In terms of carbon content, steel lies in between pure wrought iron (usually less than .01 percent carbon) and brittle pig or cast iron (3–4 percent carbon): from about 0.2 percent carbon for tough steel for machine gears or structural members, to about 1.2 percent for particularly hard steel for ball bearings or the cutting tools of your lathes. So how do you decarbonize pig iron?

The Bessemer converter is a giant pear-shaped bucket, lined with refractory bricks and mounted on pivots so it can be tipped. The vessel is charged with molten pig iron, and then air is pumped in through holes in the bottom, not unlike the action of a bubbling aquarium aerator. The excess carbon reacts with the oxygen and escapes as carbon dioxide gas, and other impurities are also oxidized and scrubbed out into the slag. One lucky outcome is that as the carbon burns it releases enough heat to keep the iron molten throughout.

The difficulty is that it is hard to judge the operation accurately enough to remove almost all of the carbon, but leave just under 1 percent. The trick for nailing the final composition, obvious in retrospect, is to run the conversion long enough that you are sure absolutely all of the carbon has been removed, and then simply mix exactly the final percentage you want back into the pure iron. This Bessemer process was the first method in history for cheaply mass-producing steel, and you'd want to leapfrog back to this point as quickly as you can.

GLASS

While iron and steel are the celebrated building stuff of the modern industrialized world, humble glass, so easily overlooked (or at least looked through), has also been critical in our progression. Glass, one of the first synthetic materials to be made by humankind, was invented in Mesopotamia, the cradle of the first cities, some time in the third millennium BC. We'll see in a bit how glass, and its unique combination of essential properties, forms the linchpin of science. But let's start with the fundamentals of how to make the stuff.

You probably know that glass is made from molten sand, or more precisely from purified silica (silicon dioxide). But simply chucking handfuls of sand into a fire won't yield you any results, other than perhaps snuffing your fire. The problem is that silica has an exceedingly high melting point, at around 1,650°C. This is far beyond the capability of a simple kiln, and so just knowing the main constituent in glass doesn't help you actually make it. Glass is sometimes produced naturally: if you dug around in the sand of a desert you might be lucky enough to unearth curious long hollow tubes of fused silica, often resembling the complex branched root system of a tree. A structure like this is called a "fulgurite," or "thunderbolt stone," and is created when lightning strikes dry sand. The electrical current surges underground and produces temperatures high enough to fuse silica grains together into a glassy tube.

Since you can't directly harness the power of lightning, in order to manufacture glass you must be able to lower the melting point of silica to within the reach of a kiln, using a suitable flux additive. Either potash or soda ash works perfectly well as a flux for silica in glassmaking, but as we'll see in Chapter 11, with a little application of chemistry, soda is much easier to produce in bulk. So the vast majority of glass made today for windowpanes or bottles is soda-lime glass—a solution of soda and lime dissolved in sand that freezes at everyday temperatures.

A ceramic crucible, fashioned from fired clay, is filled with silica grains and your soda crystals. In the heat of the kiln, the sodium carbonate decomposes (giving off carbon dioxide) and dissolves in the silica, lowering its melting point enough to successfully produce glass at the temperature of the kiln. The carbon dioxide given off, combined with oxygen and nitrogen trapped in the initial mixture, forms a bubbly, frothy melt. So a very hot kiln that keeps the molten glass very runny should be used and the crucible left inside long enough to allow these bubbles to escape and produce a clear glass. Unfortunately, glass manufactured from just silica and flux dissolves in water, severely limiting its usefulness. The solution is to use a second additive in the crucible to render the glass insoluble: quicklime—the calcium oxide we encountered in the previous chapter—works well for this.

Silica, the base material for glass, makes up more than 40 percent of the Earth's mantle and crust—it is by far the most abundant compound in the rocks of our planet. But silica is often mixed in with many other things (including metals—silica is the main component of the slag thrown away after smelting), and to make a clear, colorless glass you want it to be as pure as possible. The brownish tint to much sand, for example, is due to iron oxides, and these will tint the resultant glass green—fine for a wine bottle, irritating for a window or telescope. The best source for clear glass is a bright white sand, or other uncontaminated silica like the white quartz pebbles used for the famous Venetian "crystal" glass or the flints picked out of chalk for English "lead crystal" glass (both are technically misnomers, as all glass has its atoms arranged in a totally disordered, noncrystalline mess.)

Of course, there will be an enormous amount of glass left behind by the old civilization that can be salvaged. That which has been preserved whole can be reused, and smashed glass can be cleaned and remelted. Indeed, glass is one of the most easily recyclable materials today. It is simply melted in a furnace and then reformed, and this can be repeated time and time again without any deterioration of the material (unlike plastics, for example). But you'll need to know the necessary recipe for making glass from scratch later in the civilization-recovery process, or if you're shipwrecked on a deserted island. In fact, a tropical beach might be pretty much the ideal location to collect the three raw materials needed for clear, high-quality glass: iron-free bright white sand, seaweed to extract soda ash, and seashells or coral to calcine into quicklime.

Glass in its molten state can be poured straight out of the crucible and into a mold. But a far more useful manufacturing process exploits one of its curious attributes. Glass has the unusual property that it possesses no single melting point. Instead, the viscosity of glass (or its runniness) varies greatly over a range of temperatures, and so you can work with the material when it's in a sweet spot of being pliable but not too runny—and this allows for glassblowing. Daub a glob onto the end of a clay or long metal pipe and you can force air in to inflate the glass, either turning it in open air to work into the desired shape, or blowing into a mold to rapidly manufacture objects like bottles.

Windows are vital today for illuminating our homes and skyscrapers, allowing sunlight to flood into our artificial caves while providing a barrier to keep out the elements. The Romans were the first to glaze their windows, using small pieces of cast glass, while by the end of the first millennium AD, Chinese windows were still covered with paper, made translucent with oil. For many centuries, windowpanes were first blown and then spun flat while still soft—the distinctive dimple in the middle, still seen in the windows of old country houses or pubs, is where the glassblower's pipe was detached. Today, large, perfectly smooth window panes are made by pouring glass onto a bath of molten tin, where it floats and spreads out into a slick of uniform thickness before cooling and setting. But beyond windows, glass will offer other fundamental uses during recovery in the post-apocalyptic world.

The primary attribute of glass that makes it such a handy material for windows is, of course, that it is transparent. This in itself is a rare material property. But, in fact, glass offers a whole combination of crucial attributes that are not found in any other substance. This means that glass is utterly critical for science: studying natural phenomena, measuring their effects, and devising ever more capable technology from this knowledge. The barometer (a pressure indicator) and thermometer, for example, were the first two scientific instruments invented, and both operate by displaying changes in the level of a column of liquid. It would be impossible to see these fluctuations without the clear, hard material that is glass.

Microscope slides, too, rely on the fact that thin samples can be stuck to a substrate that lets light through. Glass is also pretty strong, and can make airtight enclosures capable of containing a vacuum. Vacuum tubes are required for the generation of X-rays (see Chapter 7 on medicine), and were critical to the discovery of electrons and other subatomic particles. Airtight glass bubbles are also central to the working of a filament light bulb or fluorescent lamp: holding a particular internal atmosphere while allowing the generated light to shine out.

As well as being transparent, heat-resistant, and strong enough to make thin-walled vessels, glass is also largely inert. And this has been critical for every aspect of chemistry research. Glass can be molded or blown into all the varied forms of laboratory apparatus: test tubes, flasks, beakers, burettes, pipettes, pipes, condensers, fractionating columns, gas syringes, measuring cylinders, and watch glasses. It's hard to see how chemistry could ever have progressed without access to a material that is both inert and see-through, allowing us to watch what happens in a reaction without tainting it.

But perhaps the greatest gift of glass is that it can be used to control and manipulate light itself. And this allows us not only to contain little pockets of nature to study in isolation, but to extend our very senses.

The Romans were masters of glassmaking, and noticed that a glass globe appeared to magnify objects behind it. But they never made the next conceptual step: to grind a lump of glass into a curved shape, creating a lens. A lens relies on the principle of refraction, where the path of a light ray is bent as it passes from one transparent medium to another. This can be seen by poking a straight stick into a pond—the stick appears to be kinked below the waterline. This is due to light rays refracting at the lake's surface, the interface between the water and air. A piece of glass formed into a particular shape, a lenticular form with a bulging, bowl-shaped curve on both sides—symmetrically convex—controls the refraction of light rays passing through it. Light arriving near the outer rim of the lens is deflected inward greatly because it hits the surface at a large angle; light passing nearer to the center is bent to a lesser degree; and rays lancing right through the middle of the lens hit its curved surface head-on and so continue straight. The outcome of this is that you have brought all the light paths together at a single point, the focus. This is the principle of a magnifying glass.

The first optical technology was spectacles, appearing in Italy around 1285 with convex lenses to help people with farsightedness, which often develops later in life, when your eyes struggle to focus on nearby objects. Correction of nearsightedness demands a concave lens, and correctly grinding a puck of glass in this opposite way—so that its two faces curve in toward the middle and light rays are instead diverged—is a little trickier.

The real breakthrough comes with the realization that if lenses can apparently enlarge objects viewed through them, a carefully arranged combination of lenses can allow you to see far into the distance—the essence of the telescope. This gadget was first used by ship captains, but was soon pointed toward the heavens to initiate the great revolution in our comprehension of the cosmos and our place within it. But glass lenses also allow you to magnify the very small, and the microscope is absolutely indispensable for understanding microbiology and germ theory, examining the structure of crystals and minerals, and improving metallurgy.

One of the first artificial substances synthesized by humanity more than 5,500 years ago, glass has enabled us to investigate nature and construct new technologies, from the first pair of reading spectacles to the Hubble Space Telescope. Of the six instruments crucial to the development of the modern scientific enterprise in the seventeenth century, all of which would be indispensible in rediscovering the world after an apocalypse—pendulum clock, thermometer, barometer, telescope, microscope, and vacuum chamber with air pump—all but one of them, the pendulum clock, rely utterly on the unique combination of properties offered by glass.

It's astounding to think that the telescopes extending our eyesight throughout the cosmos and the microscopes exploring the minute structure of matter all come down to a simple curved lump of sand. Glass, in a very literal sense, changed our view of the world. It will be crucial for a successful recovery of civilization after the Fall, both as a building material and as a critical gateway technology for conducting science. The thermometer, barometer, and microscope are all also crucial for examining the state of the human body, and so it is to medicine that we will now turn.

CHAPTER 7

MEDICINE

The city was desolate. No remnant of this race
hangs around the ruins, with traditions handed
down from father to son and from generation to
generation. . . . Here were the remains of a
cultivated, polished, and peculiar people, who
had passed through all the stages incident to
the rise and fall of nations; reached their golden
age, and perished. . . . In the romance of the
world's history nothing ever impressed me
more forcibly than the spectacle of this once
great and lovely city, overturned, desolate, and
lost. . . . overgrown with trees for miles around,
and without even a name to distinguish it.

JOHN LLOYD STEPHENS, *explorer who discovered
the remains of the Mayan civilization* (ca. 1841)

AFTER THE COLLAPSE of technological civilization, you would see the
almost complete unraveling of modern medical capability. For people
used to living in developed nations where an ambulance can be sum-
moned with a phone call, the evaporation of health care, and the loss
of the peace of mind that used to come with it, would be pretty terri-
fying. Every injury is now potentially fatal. A compound fracture of
the leg, caused by tripping over some rubble in an abandoned city, is
lethal if it doesn't receive adequate medical attention. Even an utterly

trivial incident could end up being a death sentence: a pricked finger that becomes infected and poisons the blood. So in the immediate aftermath of the catastrophe you may find a continued decline in numbers, simply because the death rate from injury and disease exceeds the birthrate. Without access to antibiotics, surgical procedures, or medications for prolonging the deteriorating body in old age, survivors can anticipate their life expectancy to plummet from the 75–80 years reasonable in the developed world today. Even if plenty of nurses, doctors, and surgeons survive, their detailed knowledge and skills will become rapidly useless without access to diagnostic equipment and blood tests or the availability of modern pharmaceutical drugs. And what if this highly specialized medical learning is itself subsequently lost? How can you accelerate the recovery of centuries of know-how?

As with most of the other topics covered in this book, it would be impossible to meaningfully describe even a minute sliver of current medical knowledge: the complex system of organs, tissues, and molecular mechanisms running the healthy human body and how they are perturbed by particular diseases or injuries; the cornucopia of pharmaceuticals we use today and how to synthesize them; or the myriad intricate surgical procedures. But what we can hope to achieve is to explain the most fundamental knowledge that will give you a fighting chance in the immediate aftermath, and describe the tools and techniques that will be essential in accelerating the rediscovery of everything else from the ground up.

Today, most of us in the West will eventually succumb to chronic diseases such as heart disease or cancer as the body starts malfunctioning with age; but, as throughout our history and in developing nations to this day, in a post-apocalyptic world it is infectious contagions that will return as the scourge of humanity.

Indeed, many of these infectious diseases are a direct consequence of civilization itself. In particular, the domestication of animals, and living with them in close proximity, allowed diseases to jump the species barrier and infect humans. Cattle transferred tuberculosis and smallpox into the human pathogen pool, horses gave us rhinovirus (the common cold), measles came from dogs and cattle, and pigs and poultry still pass us their influenzas. In addition, city living positively encourages disease: tightly packed populations allow rapid propagation of contact-based or airborne contagions, and poor sanitation and squalid conditions result in pandemics of waterborne disease. Until relatively recently, urban death rates were so high that the population of cities was maintained only by a constant influx of migrants from the countryside. But despite its risks, living together also promotes trade and the rapid transmission of far more important commodities: ideas. As the population recovers after the apocalypse, urbanization will once again foster collaboration and inspiration between people with different skill sets and specialties and will greatly accelerate the redevelopment of technological sophistication.

So let's look first at how to keep the surviving society healthy and shielded from disease, as well as ensuring safe childbirth to help the post-apocalyptic population increase as quickly as possible.

INFECTIOUS DISEASES

It would be ironic if you were fortunate enough to survive the end of the world as we know it only to perish a few months later from an easily preventable infection. In a post-apocalyptic world without antibiotics or antivirals, you desperately want to avoid becoming infected. Contagions are caused by the overwhelming of the body's defenses by microbial invaders, and understanding basic sanitation and hygiene will do more than any other single piece of information to save your life in the immediate aftermath.

We now understand well the mechanism of cholera. The *Vibrio* bacterium multiplies rapidly in the nutrient-rich soup of the small intestine, hitting the intestinal wall with a targeted molecular toxin that triggers diarrhea and aids the organism's spread to new hosts. Many enteric infections have a similar modus operandi and are spread readily by what doctors delightfully term fecal-oral transfer. The simple preventative trick is to break this cycle.

On an individual level, the single most effective thing you can do to protect yourself from potentially life-threatening disease and parasites is to regularly wash your hands (using the soap we've learned to make in Chapter 5). This isn't some ritualistic hangover from modern civilization, a matter of good manners to keep your mitts looking nice, but a basic survival skill—do-it-yourself health care. Alongside this, as a society you need to ensure that your drinking water isn't contaminated with your own or anyone else's excrement. These are the central tenets of modern public health, and retaining the most basic principles of germ theory will keep the post-apocalyptic society healthier than that of our ancestors even as late as the 1850s.

If you do succumb to an enteric infection, the good news is that the condition is often entirely survivable. Even something as historically devastating as cholera is not actually directly lethal: you die from rapid dehydration resulting from the profuse diarrhea, losing as much as 20 liters of body fluid a day. The treatment, therefore, is astoundingly straightforward, even though it was not widely adopted until the 1970s. Oral rehydration therapy (ORT) consists of no more than a liter of clean water with a tablespoon of salt and three tablespoons of sugar stirred in, to replace not only the water lost in the sickness, but also your body's osmolytes. To survive cholera you don't need advanced pharmaceuticals, just attentive nursing.

Without modern medical intervention, childbirth will once again become a dangerous time for both mother and child. Today, serious complications during birth are often resolved with a Cesarean section: the surgeon slicing through the muscular abdominal wall and into the womb to lift out the baby. Although this is now a routine occurrence, and even requested by mothers without any medical necessity, for centuries C-sections were only ever attempted as a last resort in an effort to save the child after the mother had already died or was beyond hope. The first known cases of the woman actually surviving the surgery did not occur until the 1790s, and the death rate in the 1860s was still over 80 percent. A C-section is still a very complicated and traumatic procedure today, and in the aftermath of the Fall will not soon offer any safe alternative to natural birth.

A nonsurgical way to help a baby through a difficult birth was developed sometime in the early 1600s. The birthing forceps represent a profound improvement in obstetrics, enabling the midwife or doctor to reach up the birth canal to firmly but carefully grip around the skull of the fetus and realign the head or gently pull the whole baby out.[1] An important improvement was for the two arms of the tool to be detachable off the pivot so they could be independently slid into position, and over time the design has gradually evolved for the arms of the forceps to follow the anatomical curvature of the mother's pelvis (so that the instrument works in tandem with the muscular contractions) and the

---

[1] Birthing forceps were kept a strict secret for more than a century by the family of doctors who invented them, as so much money could be made from the advantage they provided over other obstetricians. To preserve their mystery, they were brought into the room inside a lined box that was opened only when observers had been removed and the mother blindfolded.

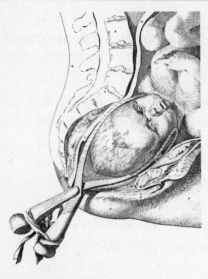

BIRTHING FORCEPS.

clamp ends to be shaped around the baby's skull.

Premature and low-birth-weight babies are likely to die if they are not kept warm in a hospital incubator until they are able to better regulate their own body temperature. Modern incubators are expensive and sophisticated machines, and like many other items of medical equipment, when donated to hospitals in the developing world today they often soon fall out of service due to power surges, unavailability of spare parts, or lack of specialist technicians to repair them; some studies find that up to 95 percent of medical equipment donated to some hospitals is inoperative within the first five years. A company called Design that Matters is attempting to address this issue, and their ingenious solution is a great example of the sort of appropriate technology that would need to emerge in a post-apocalyptic scenario. Their incubator design uses standard automobile parts: common sealed-beam headlights are used as the heating elements, a dashboard fan circulates filtered air, a door chime sounds alarms, and a motorcycle battery provides the backup electrical supply during power outages or when the incubator is being transported. All of these parts will be readily scavengeable in the aftermath and can be repaired with the know-how of local mechanics.

MEDICAL EXAMINATION AND DIAGNOSIS

The key skill of a doctor is diagnosis—being able to identify the disease or condition that a patient is suffering from, and so to determine the appropriate course of treatment or surgical procedure. The doctor asks patients to describe the details of the onset and background of their complaints and the sets of symptoms they experience. This information is then combined with the signs discovered during physical examination. The likely causes of a complaint suggested by this process help the clinician decide what follow-up investigations to request, such as blood tests, microscopic examination of samples taken from the body, or internal imaging techniques like X-rays or CT scans. The results from these investigative efforts provide the clues for reaching a diagnosis.

After the apocalypse not only will you lose the advanced tests and scanning equipment, but much of the medical expertise itself will also be lost. Medicine and surgery, more than many other areas covered in this book, rely heavily on implicit or tacit knowledge—something you have learned how to do but would find extremely difficult to successfully convey to someone else in just words or pictures. In Britain, it takes up to a decade of medical school and on-the-job learning in a hospital to achieve competency as a registrar doctor (the equivalent of a US fellow in a subspecialty), all of this with training and hands-on demonstrations provided by someone already proficient. If this cycle of knowledge transfer breaks with the collapse of civilization, it will be impossible to teach yourself the necessary practical skills and interpretative expertise from textbooks alone. So let's look at the very fundamentals of medicine and surgery: if all of the specialist understanding and equipment have disappeared, how can you recover the essential knowledge and skills?

Informed diagnosis relies on a variety of investigations, but until the early nineteenth century the medical profession didn't possess a single instrument that allowed doctors to assess the internal state of the body; they had to rely on visible external signs, prodding with their fingertips for enlarged organs or masses, or tapping the abdomen and thorax for the differing sounds of underlying air or fluid. (This percussive technique was invented by a physician who was the son of a innkeeper; he is said to have gotten the idea from the method of judging the remaining level of wine in a cask.)

The tool that transformed medical diagnosis is astonishingly simple. The stethoscope need be no more than a hollow wooden tube held to the ear and pushed against the patient's body, or even a rolled-up bundle of papers, which was how the tool was invented in 1816. René Laennec was uneasy about his ear and cheek touching the chest of a particularly buxom woman and so he improvised, realizing that the makeshift tube was not only perfectly adequate for transferring the sounds of the heart, but actually served to amplify them. A stethoscope can reveal the internal sounds of the body: not only anomalies in the sound of the heartbeat, but the wheeziness or crackling indicative of lung disease, the silence at the point of an obstructed bowel, or the faint heartbeat of a fetus.

Before the end of the nineteenth century, not only the stethoscope, but compact thermometers able to measure body temperature and inflatable cuffs linked to a gauge for measuring blood pressure, were standard items in a doctor's kit bag. The clinical thermometer can reveal a fever indicative of infection, and the pattern shown by regular readings plotted on a temperature chart can even be suggestive of certain diseases. But the stethoscope will remain your key tool for assessing the internal condition of the human body until the post-apocalyptic civilization has relearned how to generate a very high-energy form of light. Here's how.

In the closing decades of the nineteenth century, two curious emanations were discovered. The first of these was found to stream off the negative electrode when a high voltage was applied between two metal plates. These emissions were named cathode rays, and we now identify them as electrons: the agents of electric current in a wire that are accelerated away down the steep electric field created by the voltage. Flying electrons are rapidly absorbed by even tenuous matter like air, and so these cathode rays can travel an appreciable distance only inside a container evacuated of gas. Cathode rays could be noticed, therefore, only once scientists were able to produce effective vacuum pumps to suck out practically all the air in sealed glass canisters.

The small amount of gas left inside these early vacuum tubes produced an eerie glow as it was struck by the fast-moving electrons (an effect exploited in neon lights). The German physicist Wilhelm Röntgen wanted to exclude this light so that he could study the cathode rays penetrating through the wall of the vacuum tube, so he wrapped the tube in black cardboard. It was at this point that he noticed a fluorescent screen on the other side of the lab bench glowing a faint green. This was far too distant for cathode rays to be reaching, and Röntgen nicknamed this invisible new radiation X-rays, after their mysterious nature. We now know these X-rays to be ultra-high-energy electromagnetic waves emitted when the accelerated electrons slam into the positive electrode in the vacuum tube.

To his utter astonishment, Röntgen realized that X-rays allow you to see right through solid objects, such as the contents of closed wooden boxes; most eerily of all, in 1895 he was able to use X-rays to take a photograph of the bones inside his wife's hand. As X-rays are absorbed more easily by dense internal structures like bones than by soft tissues, the image essentially showed the shadow of her bones from energetic light shone straight through her body. X-rays are dangerous, as they are energetic enough to trigger mutations and cause cancer, and so patients should be exposed for only a short burst to capture a snapshot on photographic film, with the doctors shielded behind a lead screen. Despite these health risks, the opportunity offered by radiography to peer inside the living body in order to examine the vital organs, assess bone fractures, or locate tumors provides vastly greater capability for diagnosis than the first diagnostic tool, the stethoscope.

But being able to externally sense the interior condition of the body is only half of the problem you'll face after the Fall. It is absolutely crucial to be able to link patient examinations to an accurate understanding of how our body is actually built—to literally know ourselves inside out. So if this detailed knowledge of the intricacies of our own inner structure is lost, how could you rediscover it from scratch, and so recognize what is healthy and what is abnormal?

The internal construction of animals is familiar from butchery, but the human body has important structural differences. and so getting reacquainted with anatomy, gained through human dissection, will be imperative during the reboot. Anatomy and postmortem dissection will be crucial for the redevelopment of pathology—the understanding of the root causes of diseases. The practice of conducting a postmortem is absolutely vital in correlating the external signs and symptoms of sickness in the patient while alive with internal anatomical faults or defects that can be assessed only after death. The recognition that a particular disease is often caused by a problem in a specific organ, rather than being a systemic issue—as suggested by the premodern belief in the imbalance of bodily humors: blood, phlegm, and black and yellow bile—is pivotal to pathology, and this realization is in turn crucial for our ability to address the underlying cause of a disease, rather than simply trying to treat the symptoms that are manifested.

Once the fundamental cause has been identified, the next step is the prescription of medication or the undertaking of surgical intervention.

MEDICINES

Arriving at the correct diagnosis of a disease is useful only if you've already developed a set of pharmaceutical preparations that are known to be effective against particular ailments. For much of human history this has been a real stumbling block, and before the twentieth century the doctor's medicine bag was largely ineffectual: imagine the frustration at understanding the diseases killing your patients but being powerless to stop them.

Many modern drugs and treatments derive from plants, and the traditions and folklore of herbal medicine are as old as civilization itself. Almost 2,500 years ago Hippocrates—renowned for the Hippocratic oath of the physician's ethical code—recommended chewing willow to alleviate pain, and ancient Chinese herbal medicine similarly prescribes willow bark to control fever. The essential oil extracted from lavender has antiseptic and anti-inflammatory properties and is therefore useful as an external balm for cuts and bruises, whereas tea tree oil has been used traditionally for its antiseptic and antifungal action. Digitalin is extracted from foxgloves and can slow down the heart rate of those suffering from a fast irregular pulse, while the bark of the cinchona tree contains the antimalarial drug quinine, which gives tonic water its characteristic bitter flavor (and led to the British colonial penchant for sipping gin and tonics).

One particular class of drugs we'll linger on for a moment are those used for pain relief, or analgesia. While these pharmaceuticals are palliative, targeting the symptom rather than the cause, they are the most commonly taken drugs in the world, for everything from everyday discomforts like headache to more serious injuries. Analgesia is an essential prerequisite for the redevelopment of surgery. Limited pain relief can be achieved by chewing willow bark, and topical analgesia, suitable for superficial injuries or minor surgical procedures such as lancing boils, is provided by chili peppers. The capsaicin molecule that gives chilies their illusory fiery burn in the mouth is known as a counterstimulant, and, like the contrary cooling effect of menthol from mint plants, can be rubbed onto the skin to mask pain signals (both capsaicin and menthol are used in muscle-easing heat patches or ointments like Tiger Balm).

But the universal painkiller, used since antiquity, is provided by the poppy. Opium is the name of the milky pink sap that can be harvested from the poppy after it has flowered, and it has considerable pain-relieving qualities. Traditionally, opium is collected daily by making several shallow slices in the swollen, golf-ball-size seedpod of the poppy plant, allowing the sap to seep out and dry to a black latex encrustation that is scraped off the following morning. Morphine and codeine are the major narcotics in opium: the dried sap can contain up to 20 percent morphine. These opiates are far more soluble in ethanol than in water, and a potent (but addictive) tincture of opium, laudanum, is made by dissolving powdered opium in alcohol. A much less labor-intensive system developed in the 1930s uses several washes of water (often slightly acidic to improve solubility) to extract opiates from the poppy after the plant has been reaped, threshed, and winnowed—the poppy seeds kept for eating or replanting, just as you would do with cereals. In fact, 90 percent of medical opiates today are still harvested from poppy straw.

The risk, though, in taking crude decoctions or tinctures of plant extracts is that, without the capability for chemical analysis, you don't know the actual concentration of the active ingredient, and taking too much can be dangerous (particularly if, like digitalin, it interferes with your heart rate). You may have a narrow window of opportunity in the dosage: trying to hit the sweet spot of administering enough to be effective, but not so much as to become lethal.

For the vast majority of serious and ultimately fatal conditions, from pervasive infection and septicemia to cancer, no effective treatment at all is available from simple herbal concoctions. The key enabling technology that started the phenomenal medical revolution after the Second World War was prowess in organic chemistry for isolating and manipulating pharmaceutical compounds. Pharmaceuticals today are available in precisely known concentrations, and either have been synthesized artificially, or a plant extract has been modified using organic chemistry to increase the potency or decrease the side effects of the compound. For example, a relatively simple chemical modification is made to the active ingredient in willow bark, salicylic acid, that allows it to retain its efficacy as a fever-reducing painkiller but reduces the side effect of stomach irritation. The result is aspirin, the most widely used drug in history.

The key practice in evidence-based medicine that you'll need to return to after the Fall is running a fair test to see if a particular compound or treatment actually works[2]—or whether it should be thrown out alongside useless snake oils, witch-doctor potions, and homeopathic concoctions. Ideally, objectively testing a treatment's effectiveness in a clinical trial involves a meaningfully large number of patients split into two groups: one to receive the putative therapy and the other, the control group that forms the baseline for comparison, to be given a placebo or the current best drug. The two pillars of successful clinical trials are the random assignment of test subjects to the groups so as to remove bias, and the use of "double blinding": neither patients nor practitioners know who has been assigned to which group until the results are analyzed. During the redevelopment of medical science after a Fall there will be no shortcuts for meticulous, methodical work, which may also call for disagreeable practices like animal testing for the sake of easing human suffering.

---

2 One of the first clinical trials in history was conducted in 1747 with scurvy sufferers to demonstrate that citrus fruits do in fact contain a protective agent.

For some conditions, the best course of action is surgery: to physically correct or remove the faulty or troublesome component of the body's machinery. But before you can even think about attempting surgery (with a reliable chance of patient survival)—intentionally creating a wound to open the body, having a look inside, and tinkering with the workings inside like a car mechanic—there are several prerequisites that a post-apocalyptic society will need to develop. These are the three As: anatomy, asepsis, anesthesia.

We have already seen that you need to know how our body is built so that you can tell a diseased organ from a healthy one. And without a detailed grasp of anatomy, your surgeons are literally poking around in the dark. You need to have a comprehensive map of the body's internal makeup, the normal forms and structures of each of its components; you need to understand their function and know the paths of major blood vessels and nerves so that you don't accidentally sever them.

Asepsis is the principle of preventing microbes getting into the body in the first place during surgery, rather than trying to clean the wound later with antiseptics like iodine or ethanol solution (antiseptics are your only option for an accidental, dirty wound). To maintain aseptic conditions, scrupulously clean the operating theater and filter the air supply. The site of the operation can be cleansed with 70 percent ethanol solution before the incision is made, and the patient's body covered with sterile drapes. Surgeons must wear clean surgical gowns and face masks, scrub their hands and forearms, and operate with heat-sterilized surgical instruments.

The third crucial element is anesthetics. These are drugs that don't cure disease but do something just as valuable: they can temporarily pause all sensitivity to pain, or even induce complete unconsciousness. Without this, surgery is an abominably traumatic experience and can be attempted only as a last resort. The surgeon must work rapidly, slicing through muscular tensions and spasms as the patient writhes in agony, and only simple procedures can be considered: removal of a kidney stone or the brutish amputation of a gangrenous limb with a butcher's saw. With an insensate patient, however, surgeons can afford to work much more slowly and carefully, and are able to risk invasive operations on the chest and abdomen, as well as exploratory surgery to see what might be the underlying causes of an ailment.

The first gas to be recognized for its anesthetic properties was nitrous oxide, or "laughing gas": when it is inhaled at high enough doses, its exhilarating sensation can give way to true unconsciousness, suitable for surgery or dental work. Nitrous oxide is generated from the decomposition of ammonium nitrate as it is heated—be careful, though, as the compound is unstable and may explode if it gets much hotter than 240°C—and the anesthetic gas is then cooled and cleaned of other impurities by bubbling through water. Ammonium nitrate can itself be produced by reacting ammonia and nitric acid (see Chapter 11). Nitrous oxide alone is good for dulling the sensation of pain, but it is not very powerful as an anesthetic. If, however, it is administered with other anesthetics, such as diethyl ether (often abbreviated to ether), it acts to potentiate them, enhancing their effectiveness. Ether can be produced by mixing ethanol with a strong acid, such as sulfuric acid, and then extracting ether from the reaction mixture by distillation. It is a reliable inhalation anesthetic, and although ether is relatively slow to work and can produce nausea, it is medically safe (although it is an explosive gas). The advantage of ether is that not only does it induce unconsciousness, but it also acts to relax muscles during surgery and provides pain relief.

But what if, generations after the Fall, society has regressed so much that the vital knowledge of germ theory has been lost, and pestilence is once again attributed to bad air (*mal aria*) or fractious gods? How could a post-apocalyptic civilization rediscover the existence of unimaginably tiny creatures invisible to the eye that cause food spoilage, festering wounds, putrefaction of corpses, and infectious diseases?

In fact, bacteria and other single-celled parasites can be seen with some beguilingly simple equipment. A rudimentary microscope is surprisingly easy to make from scratch. You'll need to start with some good-quality, clear glass. Heat the glass and draw it out into a thin strand, and then melt the tip of this in a hot flame so that it drips. The globule cools as it falls, and with luck you'll produce some very tiny glass beads, perfectly spherical in shape. Use a thin strip of metal or cardboard with a hole in the middle to mount your spherical lens, and hold it over a sample. This simple microscope works because the tiny ball of glass has a very tight spherical curvature and thus a powerful focusing effect on light waves passing through it. This also means that the focal length is exceedingly short, though, and you will need to position the lens and your eyeball right down close to the target.[3]

---

3 Using this design, in 1681 Antoni van Leeuwenhoek became the first person in history to actually see a germ. Leeuwenhoek had come down with a case of the runs and felt compelled to examine his own watery waste under his new microscope. He reported sighting "animalcules a-moving very prettily," "somewhat longer than broad, and their belly . . . furnished with sundry little paws." He saw what we would now identify as a protozoan called Giardia, a common cause of diarrhea. Before too long Leeuwenhoek was observing microbes in droplets of water, and bacteria swarming in feces and decaying teeth. On examination of his own semen he discovered the vigorously wriggling sperm cells behind the sexual reproduction of all animals (although he insisted that he did not obtain his own samples by "any sinful contrivance" but that they were the "excess with which nature provided me in my conjugal relations").

The realization born of your instrument-enhanced senses is that there's a whole teeming universe of invisibly small organisms down there—astonishingly diverse varieties of new wildlife for post-apocalyptic micro-naturalists to identify and sort into related families and groups. With the rigor demanded of scientific proof, you can demonstrate not only that microbes are present in infected wounds or spoiled milk, but that food is preserved if microbes are *not present*. If you seal nutritious broth or corruptible meat within an airtight jar and heat it to inactivate any microbes already present, no decomposition will occur: things don't spoil spontaneously. Better microscopes can be constructed, similar to a telescope, from combinations of lenses, and in time you'll be able link the presence of specific microorganisms to particular infectious diseases.[4]

You can even grow and study these microorganisms in captivity, culturing them in flasks of liquid broth or as colonies on the surface of a solid nutrient. Petri dishes can be molded from glass, filled with nutrient-enriched agar poured in to set, and fitted with a lid to prevent contamination. Agar is a gel-forming substance extracted from boiled red algae or seaweed (and common in Asian cuisine), similar to the gelatin derived from cattle bones but indigestible by most microbes.

In earlier chapters we have seen that this fundamental microbiology is needed for the optimization of processes such as making leavened bread, brewing beer, preserving food, and producing acetone. But perhaps most important in improving the human condition after the

---

4 The possibility of invisibly small organisms was speculated upon long before the first microscope was invented. The Roman author Marcus Terentius Varro expressed in 36 BC his belief that "there are bred certain minute creatures that cannot be seen by the eyes, which float in the air and enter the body through the mouth and nose and there cause serious diseases." History might have played out very differently indeed if Varro had known how to make a rudimentary glass-globule microscope to confirm his hunch. Imagine the pestilence and suffering that could have been prevented if germ theory had been developed before the birth of Christ.

Fall, microbiology provides the prerequisite knowledge base for discovering more targeted methods than noxious antiseptic chemicals for killing bacteria and curing infection.

In 1928 Alexander Fleming had been working on cultures of *Staphylococcus aureus* bacteria from skin abscesses before leaving for a holiday. On his return he started clearing his lab bench and washing up the old Petri dishes. Randomly picking up one from the top of a pile in the sink that had not yet been treated with disinfectant, he noticed a small patch of mold surrounded by a ring clear of bacteria on an otherwise overgrown plate. It seemed that some substance secreted by the mold, later identified as a species of *Penicillium*, had inhibited bacterial growth. Penicillin, the secreted compound, and numerous other antibiotics discovered or synthesized since, are extremely effective at treating microbial infections and save millions of lives every year.

"The most exciting phrase to hear in science, the one that heralds new discoveries," said science fiction author Isaac Asimov, "is not 'Eureka!' ["I found it!"], but 'That's funny . . .'" This is certainly true of Fleming's chance finding, along with many other serendipitous discoveries, but only if the implications are grasped. Indeed, fifty years earlier other microbiologists had noticed that *Penicillium* prevented bacterial growth, but had not made the next conceptual leap from this observation to pursuing the ramifications for medicine.

With hindsight, however, and knowing of the existence of such effects, could a rebooting society replicate a similar series of experiments to deliberately search for effective molds and so rapidly rediscover antibiotics? The basic microbiology is straightforward. Fill Petri dishes with a beef-extract nutrient bed that is hard-set by seaweed-derived agar, smear across *Staphylococcus* bacteria picked out of your nose, and expose different agar plates to as many sources of fungal spores as you can, such as air filters, soil samples, or decaying fruits and vegetables. After a week or two, look carefully for molds that have inhibited the growth of bacteria around them (or indeed other bacterial colonies that do so: many antibiotics are produced by bacteria locked in an evolutionary arms race with one another). Pick them off to isolate the strain and attempt to grow it in liquid broth to make the secreted antibiotic more accessible. Antibiotic screens have now found numerous compounds from fungi and bacteria, although *Penicillium* molds are so common in the environment they are likely to be among the first re-isolated after the apocalypse. They're one of the principal causes of spoiling food: in fact, the *Penicillium* strain responsible for most of the penicillin antibiotic produced worldwide today was isolated from a moldy cantaloupe in a market in Illinois.

However, even for a rough-and-ready post-apocalyptic therapy you can't simply inject the antibiotic-containing "mold juice" because, without refining, its impurities will trigger anaphylactic shock in the patient. The chemistry worked out by Howard Florey's research group at the end of the 1930s to purify penicillin from the growth medium exploits the fact that the antibiotic molecule is more soluble in organic solvents than in water. Strain the growth culture to remove bits of mold and detritus, add a little acid to this filtrate, and then mix and shake with ether (we saw earlier in this chapter how to make this versatile solvent). Much of the penicillin will pass from the watery growth fluid into the ether, which you need to let separate and rise to the top. Drain off the bottom watery layer, and then shake the ether with some alkaline water to entice the antibiotic compound to pass back into the aqueous solution, now cleansed of much of the crud in the growth fluid. The daily dose of penicillin for a single person prescribed today requires up to 2,000 liters of mold juice to be processed, and so post-apocalyptic antibiotics will demand a high level of organized effort to produce. By the end of 1941 Florey's team had scaled up production to make enough penicillin for clinical trials, but they were forced by wartime shortages of equipment to improvise. Mold cultures were grown in racks of shallow bedpans and makeshift extraction equipment built using an old bathtub, trash cans, milk churns, scavenged copper piping, and doorbells, all secured in a frame made from an oak bookcase discarded by the university library—inspiration, perhaps, for the scavenging and jury-rigging necessary after the apocalypse.

So while the discovery of penicillin is often portrayed as accidental and almost effortless, Fleming's observation was only the very first step on a long road of research and development, experimentation and optimization, to extract and purify the penicillin from the "mold juice" to create a safe and reliable pharmaceutical. In the end, the United States provided the large-scale fermentation to supply enough for widespread treatment. Similarly, once it understands the necessary science, a post-apocalyptic civilization will need to reattain a certain level of sophistication before it can produce enough antibiotic for it to have an impact across the population.

CHAPTER 8

POWER TO THE PEOPLE

The white flashed back into a red ball in the
southeast. They all knew what it was. It was
Orlando, or McCoy Base, or both. It was the power
supply for Timucuan County. Thus the lights went
out, and in that moment civilization in Fort Repose
retreated a hundred years. So ended The Day.

PAT FRANK, *Alas, Babylon* (1959)

FLICKING BACK THROUGH the gas and electricity bills for my apartment
in north London, my total energy consumption last year was a little
under 14,000 kilowatt-hours (kWh). If, without access to fossil fuels,
all of this energy were to be provided by maintained forestry, I'd need
to burn almost 3 tons of dried wood (or 1.7 tons of more-condensed
charcoal) every year, which would require more than half an acre of
short-rotation coppiced woodland. But that's assuming that it's possible to successfully convert 100 percent of the energy locked up in a log
into electricity flowing from my outlets. In fact, the multistep process
of combusting fuel to generate electricity is inherently inefficient, and
even modern power stations can convert around only 30–50 percent of
the stored energy of their fuel into electricity.

And of course, that's only counting the energy I use directly within
my four walls, for heating, lighting, and running appliances. It misses
all of that expended to support my share of the industrialized civiliza-

tion I live in—the energy used in road building and construction, the industrial processes needed to provide me with writing paper and powdered detergent, the energy required to manufacture and transport my clothes or sofa, and to synthesize fertilizer and plow fields for my meals, and the fuel burned by the train I take to work. When you divide national energy consumption by total population, you find that each individual living in the United States actually uses nearly 90,000 kWh every year, while a European uses just over 40,000 kWh.

Before the mechanical revolution in the Middle Ages that began the widespread use of waterwheels and windmills, and later, industrialization based on the exploitation of fossil fuels, the effort needed for agriculture, manufacture, and transportation was provided by muscle power alone. If we put this modern energy consumption into perspective, 90,000 kWh is equivalent to every American having a team of fourteen horses, or more than a hundred humans, working flat-out, 24/7 for them.

With the fall of industrialized civilization and the disintegration of this energy feed, the recovering post-apocalyptic society will have to relearn how to provide for its energy requirements. The advance of civilization is based on being able to marshal greater and greater energy resources, and especially on learning how to convert between energy types, gaining the capability to transform heat into mechanical power, for example.

MECHANICAL POWER

Civilization requires not just thermal energy, as we saw in Chapter 5, but also the harnessing of mechanical power, relieving it from the constraints of using muscle power alone.

One of the key Roman innovations was the development of the

vertical, geared waterwheel: the bottom of a large wheel with paddles is dipped into a stream or river and turned by the force of the flow. In antiquity, this water power was primarily applied to turning a grinding stone to mill flour, and the crucial mechanism that allowed this technology was the invention of the right-angle gear (dated to around 270 BC), transforming the direction of motion from the vertical spin of the waterwheel to the horizontal rotation of the grinding stone. Most simply, this can be achieved with a large crown wheel (one with pegs sticking out of the flat face of the gear) on the waterwheel drive shaft, coupled to a cylinder of rods known as a lantern gear or cage gear that is connected to the millstone. Altering the relative sizes of the crown wheel and lantern gear allows you to match the required speed for

OVERSHOT WATERWHEEL. THE RIGHT-ANGLE GEAR CONVERTS THE VERTICAL MOTION INTO HORIZONTAL ROTATION SUITABLE FOR DRIVING MILLSTONES TO GRIND FLOUR.

grinding to the flow rate of different rivers. These water mills were the very first known application of gearing to transfer power, and so represent the earliest roots of mechanization.

Although it can be dunked in the flow from practically any riverbank, or even mounted over the side of a milling boat anchored in the current, the undershot wheel is woefully inefficient, and in its simplest form suffers from problems with varying river levels. Luckily it doesn't take much technical know-how to build a far more capable and powerful waterwheel. The overshot wheel became widely exploited across Europe during the supposedly ignorant and stagnant "Dark Ages" following the fall of the Roman Empire, and, despite similarities in overall appearance, functions on a completely different principle than the primitive undershot wheel.

Rather than being stuck into the flow, the bottom of the overshot wheel is held clear of the tailrace, and water is delivered to the very top of the wheel by a chute. The overshot wheel derives its torque not from the impact of a current, but from the energy relinquished by the water as it falls. This design is far more efficient and can capture as much as three-quarters of the energy held in the head of water. Fit a sluice gate to the chute to control the flow onto the wheel, and if the stream is dammed to create a mill pond, a reservoir of energy can be built up until it is required to be expended (something that wasn't attempted until the sixth century AD, half a millennium after the first vertical waterwheels were used, but could be leapfrogged to during a reboot).

Harnessing wind is technically much trickier than tapping into water power, and consequently the technology arrived much later in our history of development (although boats with sails to catch the wind for propulsion date back to 3000 BC). Water is a far denser medium than air, and so even a gentle flow carries a great deal of energy, making it an easy resource to exploit even with imperfectly designed elements and inefficient wooden gearing. Unlike the sluice gate, you have

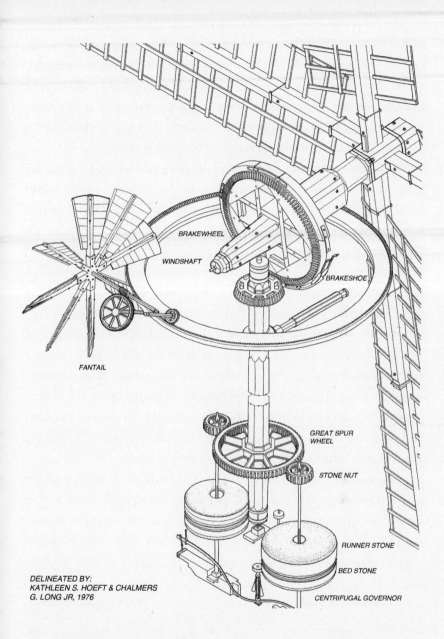

BRAKEWHEEL

WINDSHAFT

BRAKESHOE

FANTAIL

GREAT SPUR
WHEEL

STONE NUT

RUNNER STONE

BED STONE

DELINEATED BY:
KATHLEEN S. HOEFT & CHALMERS
G. LONG JR, 1976

CENTRIFUGAL GOVERNOR

SELF-ORIENTING TURRET WINDMILL. THE FANTAIL KEEPS THE MAIN SAILS
TURNED INTO THE WIND, AND THE CENTRAL SHAFT DRIVES TWO SETS OF
MILLSTONES.

no control over the strength of the wind, so if it begins blowing too briskly, the windmill blades or driven mechanisms can be damaged. Windmills therefore need a braking system and a method to control the effectiveness of the blades, such as reefing canvas sails. The most fundamental challenge, however, is the constantly changing wind direction; a windmill needs to be able to be quickly reoriented.

Rudimentary windmills can be built on a post and the entire structure manually turned to the wind, but for larger and more powerful fixed windmills the blades need to be mounted on a top turret able to automatically swivel around the central drive shaft to face into the wind. The mechanism employed here is ingeniously simple: a small fan behind, and facing at a right angle to, the main sails is geared to a toothed track running around the top rim of the tower, so that whenever the wind changes and blows across this fantail, it spins and rotates the turret around until it is oriented perfectly in line with the wind again.[1]

All of this demands a much greater degree of mechanical sophistication than even the largest waterwheel. But once you've mastered wind power, your sites of production are liberated from the watercourses and can occupy even flat landscapes (like the Netherlands), or regions either without abundant water resources (such as Spain), or that are often frozen over (like Scandinavia).

The taming of the wild power of both wind and water, coupled with the increasingly effective use of draft animals (we'll return to this later), had a profound impact on our society, and you'll want to achieve the same level as rapidly as possible during the reboot. Medieval Europe became the first civilization in human history to base its pro-

---

1 As they achieved an impressive degree of sophistication in the late nineteenth century, windmills became controlled by a centrifugal governor—two heavy balls that swing out on arms—that automatically regulated the spacing between milling stones to suit the variable wind speed. Today we instantly associate this control system with the steam engine, where it acts to close the throttle valve admitting high-pressure steam into the piston if it begins to whirl too rapidly, but James Watt had in fact borrowed it wholesale from windmill technology.

ductivity not on human muscle power—the labors of coolies or slaves—but on the exploitation of natural power sources. This mechanical revolution, gathering momentum between the eleventh and thirteenth centuries, went far beyond the use of a mill to pulverize the harvest's grain into flour. The potent torque of the waterwheel and the windmill became a ubiquitous power source for a staggeringly diverse range of applications:

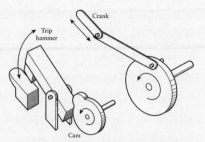

FUNDAMENTAL MECHANISMS: THE CRANK (RIGHT) TRANSFORMS ROTATION INTO A BACK-AND-FORTH MOTION SUITABLE FOR SAWING, AND THE CAM (LEFT) CAN BE USED TO REPEATEDLY LIFT AND DROP A TRIP-HAMMER.

pressing olives, linseed, or rapeseed for oil; driving wood-boring drills; polishing glass; spinning silk or cotton; powering metal rollers to squash iron bars into shape. The elementary mechanical component that is the crank arm transformed rotary motion into a reciprocating thrust suitable for mechanizing sawmills, ventilating mine shafts, or pumping water from mines or flooded lowlands (as employed to great effect by the Dutch). But perhaps the most versatile function was turning a cam to repeatedly lift and drop a trip-hammer—perfect for crushing metal ore, pounding out wrought iron, crumbling limestone for agricultural lime or mortar, beating dirty sheep wool to full it (to clean and compact it), and pounding mash for beer, pulp for paper, bark for tanning, and woad leaves for blue dye.

The cam mechanism was employed to heave trip-hammers for seven centuries before being replaced by steam-powered versions in the Industrial Revolution, but it lives on today under the hoods of our cars and trucks, opening and closing the engine valves in the correct sequence (see Chapter 9).

So with the appropriate internal mechanism to convert the principal rotation into the desired action, medieval water and windmills were the original power tools. The medieval world may not have been industrial, but it was certainly industrious. And if our civilization catastrophically collapses there is hope that this technology can be employed again to rapidly reattain a base level of productivity during the reboot.

Any civilization must successfully marshal both thermal and mechanical energy. But how do you convert between these forms? Turning mechanical energy into heat is trivial—imagine rubbing your hands together on a cold day—and indeed, trying to minimize friction and the loss of useful energy to heat is the whole point of engine lubricants and ball bearings. Being able to convert the other way, though, would be exceedingly useful. Thermal energy can be provided on demand, by burning any of a number of fuels, and the capability to transform this heat into mechanical power would release you from reliance on the vagaries of wind or water and also offer a power plant for mechanical transport. The first machine in history able to effect this transformation—to convert heat into useful motion—was the steam engine.

The central concept behind the steam engine goes all the way back to the ages-old mystery, well known to Galileo in the late 1500s, that a suction pump can't raise water more than about 10 meters up a pipe. The explanation of this is that the air itself exerts a pressure, a force squeezing everything on the Earth's surface, including the column of water. The implication is that the atmosphere itself can be made to do work for you. All you need is to create a vacuum within a smoothly bored cylinder with a freely movable piston and the air pressure outside will forcibly plunge the piston down; you can couple this to machinery for effortless labor. But how do you repeatedly generate a vacuum inside the cylinder? The answer is by using steam.

Vent hot steam from a boiler into the cylinder and then allow it to cool: as it condenses from vapor to liquid water, the pressure it exerts plummets and no longer balances that of the atmosphere. The piston is driven in by the force of the outside air, doing the work for you, and you can repeat the cycle by opening a valve to allow the piston to return, and then squirting more steam in again. This is the basic operating principle of the earliest "fire engines" of the eighteenth century, and you can make certain efficiency improvements, such as adding a separate condenser so that you're not repeatedly cooling and reheating the cylinder. But if you're able to construct sturdier cylinders and boilers, perhaps from scavenged materials or by redeveloping skill in metallurgy, you can do much better. Rather than using the sucking effect of steam condensing in the cylinder, build the steam up to a higher pressure and you can use the expansive force of the hot gas—the same whoosh as in an espresso machine—to drive the piston first one way within the cylinder, then back again from the other side.

The primary output of a steam engine (as with any piston-based heat engine, like the car motor we'll return to in Chapter 9) is the plunging back and forth of the piston. This is fine for pumping water from mines, but for most applications you'll want to transform that reciprocating motion into a smooth rotation. The crank will perform this conversion for you, just as we saw for windmills, and produce an action suitable for driving machinery or a vehicle's wheels.

You might think that steam engines represent exactly the sort of transitional technological level that you would aspire to leapfrog over during a reboot, straight to internal combustion engines or steam turbines, which we'll explore in detail later. But steam engines offer two major advantages over more advanced alternatives, and so you may need to recapitulate this developmental stage. First, they are external combustion engines and don't require refined gasoline, diesel, or natural gas to run—they are much less fussy, and their boiler can be fired with pretty much anything that burns, including scrap wood or agricultural waste. Second, a simple steam engine can be constructed with much more rudimentary machine tools and materials and with far more forgiving engineering tolerances than a more complex mechanism. We'll return to mechanical power shortly, but for now let's take a look at how to reboot one of the cardinal features of the modern world: electricity.

ELECTRICITY

Electricity, or, to be more precise, the whole set of phenomena bundled within electromagnetism, is such an important gateway technology that you would really want to beeline for it during a reboot. The discovery of electromagnetism provides a great historical example of stumbling upon a completely new field of science, offering a whole assortment of related phenomena and therefore exploitable possibilities. These novel phenomena were then mined for technological applications, which themselves in turn opened up new avenues of fundamental scientific research.

Electricity in a steady, sustained flow, suitable for exploiting for practical purposes, was first produced by the battery. A battery is actually startlingly simple to construct. All you need to create a constant electric current is two different kinds of metal, both immersed in a conducting fluid or paste called the electrolyte.[2] All metals have a particular affinity for the particles called electrons, and when two dissimilar metals are brought together, one of the pair will give them up to the more electron-hungry metal, causing a current along the wire connecting them. All batteries, whether in a mobile phone, a flashlight, or a pacemaker, encapsulate a chemical reaction that has been subjugated to run only when the connection is completed, and the flow of electrons is then channeled along a convoluted pathway of wire to do

---

2 If you have any of the old-style tooth fillings you can even demonstrate this in your own mouth. Chewing a piece of aluminum foil introduces a second metal that reacts with the mercury-silver filling in your tooth, your own saliva serving as the electrolyte. Be careful trying this, though, as the electrical current produced will be delivered right to the nerve endings in your filled teeth!

work for us. The difference in reactivity between the two metals involved determines the electrical potential, or voltage, that a battery produces.

A reasonable voltage is produced by coupling silver or copper with higher-reactivity metals such as iron or zinc. The first battery, the voltaic pile, was constructed in 1800 by stacking alternating disks of silver and zinc, separated by cardboard pads soaked with salty water. Silver, copper, and iron were all known millennia before the invention of the voltaic pile, and although zinc is harder to isolate, it is present in ancient bronze alloy and was available in pure form from the mid-1700s. Wires can be made simply by rolling or pulling soft copper. So there would seem to have been no insurmountable hurdles to electricity being discovered in classical times.

In fact, perhaps it was.

In the 1930s several curious artifacts were unearthed at an archeological dig near Baghdad in Iraq. Each is a clay jar, about 12 centimeters tall, and dated to the Parthian era (200 BC–200 AD). But it's the contents of the pottery that are so remarkable. Inside each jar is an iron bar surrounded by a sheet of copper rolled into a cylinder, and the jar shows signs of having contained an acidic fluid like vinegar. The two metal pieces are kept from touching each other, and the jar mouth is sealed with insulating natural bitumen. One hypothesis is that this ancient relic constitutes an electrochemical cell, perhaps employed for electroplating gold onto jewelry, or maybe the tingling current was believed to have medicinal properties. Replicas made of the "Baghdad battery" do indeed successfully produce around half a volt, but it's fair to say that the evidence for any electroplated items is weak, and the interpretation of these mysterious pots remains controversial. However, if it was built for the purpose of providing electricity, as is certainly possible, it would predate the voltaic pile by well over a millennium.

If the chemical reaction of stripping electrons off the negative terminal and passing them onto the positive electrode is reversible, then you have the makings of something particularly useful: a rechargeable battery. The easiest rechargeable battery to build from scratch is the lead-acid battery, common today in cars. A sheet of lead is used for each electrode, and bathed in sulfuric acid electrolyte. Both electrodes will react with the acid to make lead sulfate, but during charging you will convert the positive one to lead oxide (lead rust) and the negative one to lead metal, which is neatly reversible again as the battery discharges. Each of these cells will produce just over 2 volts, and so six of them wired together in a series gives you the 12 volts of a car battery.[3]

The problem with batteries, though, is that although they offer a fantastically portable power supply that our laptops, smartphones, and other modern gadgets rely upon, you're merely tapping the chemical energy already held in the dissimilar metals (in the same way burning a log of wood only liberates the chemical energy of carbon reacting with oxygen). You'll need to put a lot of energy into refining the reactive metals in the first place, or topping up the rechargeable battery from another electrical outlet. Batteries are a store, not a source.

The features of electricity that we rely upon so much in modern life are a related cluster of phenomena that were stumbled upon beginning in the 1820s. Place a compass next to a wire carrying a current from a battery and you'll notice that the needle is deflected. The wire is puffing out a magnetic field that locally overwhelms that of the Earth's global field, and so the compass needle reorients. You can maximize this effect by wrapping the wire into a tight coil around an iron rod core; the small fields from the wire all combine to create a powerful electromagnet that you can command on and off with the flick of a switch, and use to permanently magnetize other pieces of iron.

So if electricity can create magnetism, is the reverse also true—can a magnet conjure up a current within a wire? Indeed it can. A magnet

---

3 The root of this name for a connected assemblage of individual electrochemical cells is in military jargon: an emplacement of several heavy guns is an artillery battery.

pulled back and forth, or spun, or even an electromagnet flicked on and off, will all induce a current in a nearby coil of wire. And the faster the magnetic field moves across the wire, the greater the current induced. So electricity and magnetism are symmetrical powers inseparably intertwined with each other: two sides of the same electromagnetic coin.

And it is this simple observation of magnetism inducing current that unlocks an enormous wealth of modern technology: using a magnet, motion itself can be converted into electrical energy. You're not limited to batteries that require expensive metals and run down: you can generate as much electricity as you like from spinning a magnet inside a wire coil, or vice versa. And the converse is also true: electromagnetism can cause motion. If you place a strong magnet alongside a wire you'll notice the wire twitches as the current through it is turned on. This is the motor effect, and with a little experimentation you'll work out how to arrange the current-carrying wires and magnets (or even electromagnets) to drive a rapidly spinning shaft. Today, the electric motor drives industrial machinery, saws timber, and grinds flour, and you'll be able to count dozens of them in your home: running the vacuum cleaner, turning the exhaust fan in the bathroom, or spinning the DVD in the player. Our lives today are eased by this miniaturization of labor, with the electric motor now ubiquitous and practically invisible.

Using this principle of electromagnetism causing motion, you'll also be able to construct instruments for accurately measuring the fundamental attributes of electricity: how much current is flowing and what voltage it is at. (The earliest electricians attempted to measure it by rating the pain of the shock delivered to their tongue!) As we'll see in Chapter 13, being able to reliably quantify a new phenomenon is the critical first stage in coming to understand it and so being able to technologically harness it for your uses.

Electric light, too, has a powerful role in our modern lives, providing illumination on demand that has fundamentally changed our sleep patterns and working lives; our buildings and streets now blaze with a billion tiny suns. The simplest form of electric illumination is the arc lamp. This was invented in the early 1800s, supplied from voltaic piles, and is essentially just a continuous spark—an artificial bolt of lightning—held between two carbon electrodes. The trouble with arc light is that it is unbearably intense and thus isn't suitable for interior lighting. So while using electricity to generate light is simple, using electricity to create a practical glow is fiendishly tricky.

The physical phenomena that the light bulb is designed to exploit are simple enough. You can use the material property of electrical resistance to heat up a thin filament by pushing a current through it. As materials get hotter they begin glowing with their own light—incandescence—like an iron bar shoved into a fire, becoming cherry red, then orange, yellow, and finally a brilliant white. But the devil is in the details. If a filament of carbonized thread or metal glows white-hot in air, it rapidly reacts with oxygen and burns up. You could envelop the filament in a sealed glass orb and suck out all the air with a vacuum pump, but hot materials readily evaporate in a vacuum. Filling the bulb with an inert gas like nitrogen or argon at low pressure works well, but you will still need some R&D, trial, and error with strands of different carbonized materials or thin metal wires to find what works as a reliable filament.

GENERATION AND DISTRIBUTION

We've seen how a generator works to convert movement into electricity, but how do you create that rotation in the first place? The immediate solution is that you simply install the generator in a rudimentary windmill or waterwheel you've constructed. Generators work best spun at many hundreds of revolutions per minute, so you'll need a system of

CHARLES BRUSH'S 17-METER-DIAMETER, ELECTRICITY-GENERATING WINDMILL, BUILT 1887.

gears, or pulleys and belts, to multiply up the slow but high-torque rotation of the drive shaft. A rebooting civilization might therefore conceivably resemble a steampunk mishmash of incongruous technologies, with traditional-looking four-sail windmills or waterwheels harnessing the natural forces not to grind grain into flour or drive trip-hammers, but to generate electricity to feed into local power grids.

A feasibility study in 2005 calculated that retrofitting a single traditional four-sail windmill with a gearbox and generator to replace the millstones could produce more than 50,000 kWh of electricity a year— enough to supply my apartment four times over. But perhaps the most inspiring example of what might be achievable with rudimentary means during a post-apocalyptic reboot is offered by the American inventor Charles Francis Brush. In 1887 he built a tower on his property holding a fan 17 meters across, composed of 144 rotor blades of thin, twisted cedar wood. This could generate more than a kilowatt of electricity, which he used to power the hundred or so incandescent light bulbs—themselves cutting-edge technology at the time—throughout his mansion, with any surplus stored in more than 400 rechargeable batteries in his basement.

The problem with such designs is that the extensive system of gears needed to multiply up the slow turning wastes a lot of the energy. The solution for windmills is to fundamentally change the design. Instead of deploying broad sails that catch large amounts of the wind as it passes, but also generate lots of turbulence and drag and can therefore never spin very fast, modern wind turbines sport a triplet of long, slender blades. These are based on the lessons of aerodynamics learned from developing the propellers on aircraft, and although their much smaller surface area means they struggle to get going at slow wind speeds, they can spin incredibly rapidly in a stiffer breeze and convert far more of the rushing energy into electricity.

The power output of a waterwheel is also fundamentally limited. The amount of energy available in a stream of water is determined by the discharge and the head. Discharge is the flow rate, whereas the head of water is the total height that it drops—between the delivery chute and the race in the case of an overshot wheel. Waterwheels are severely restricted by the maximum head they can utilize, constrained by the diameter of the wheel: you can't construct a wheel much more than 20 meters across before it becomes too heavy and inefficient as it turns.

Water turbines, however, do not have the same limitation. China's Three Gorges dam, the most powerful hydroelectric plant in the world, provides a head of 80 meters between the top of the reservoir and the turbines at the base, and so can deliver prodigious energy.

A simple turbine you can build that is best at exploiting a large head

PELTON TURBINE.

and a small discharge flow (i.e., a narrow pipe producing a jet of high-pressure water) is the Pelton turbine, which consists of a ring of cups fixed around the rim of a hub (it looks a bit like a circular splay of spoons). The key is for the jet of water not to stop in each cup, but to be smartly turned around and splash out the front again. Each cup is designed like a smoothly curved bucket with two halves and a cusp-like ridge running through the middle, so that the jet of water striking the cup head-on is neatly parted by the central ridge, swirls around the curve of both halves, and streams out the front again. It is this reversal of direction that exerts a strong force on the cup and spins the turbine, the jet striking each of the cups in turn as the hub spins around.

For the opposite situation, when the available flow you have is low head but high discharge, the cross-flow turbine is better suited. Here the water is guided in through the top of a wheel with short curved vanes arranged radially, which get thrust to the side by the flow, and then again as the water exits out the bottom. The cross-flow turbine superficially resembles a traditional waterwheel, but, most important, is not turned by the weight of falling water caught in buckets but by the action of the flow of water against the backs of its curved blades.

Both the Pelton and cross-flow types of turbine are easy to construct with rudimentary metalworking tools, and today are recommended as appropriate technologies for local manufacture in the developing world. They are exactly the sort of technology that could help a rebooting post-apocalyptic society.

Despite the efficiency of wind or water turbines and their harnessing of renewable energy, most of our electricity today is not generated in this way. In fact, the age of steam never really ended. Though we don't use steam engines as prime movers for machinery or vehicles anymore, more than four-fifths of the electricity used around the world is generated using steam: firing a boiler with the heat released by combusting coal or gas, or by the disintegration of unstable heavy atoms in a nuclear fission reactor.

As we've explored, producing heat is straightforward, but transforming thermal energy into movement is a trickier step. A steam engine will do that for you, but the slow thrusting of the piston cannot efficiently be converted into the rapid rotation suitable for an electrical generator.

The solution is the steam turbine, based on successful designs for water turbines but optimized for high-pressure steam. Power can be extracted from the rush of steam either by catching the flow on the back face of the blades so that they are pushed by the impulse (like a Pelton or cross-flow water turbine), or by deflecting the water over a curved surface so that it is pulled forward by the reaction force, like an aircraft wing. The key difference from water is that steam expands, to rush faster but at lower pressure, so most steam turbines combine a reaction stage for the high-pressure steam with impulse rotors farther down the shaft when the steam has expanded. It is this multistage steam turbine that has enabled the generation of prodigious quantities of electricity with very high efficiency, and thus ushered in the modern electrical age.

For the generated electricity to be useful, though, you've got to be able to distribute it to where it's needed.

Although you can rig up a generator to provide a steady direct current (DC, same as a battery), it's easier to build one that produces a rapidly cycling alternating current (AC) as the rotor spins. The generated voltage in the coil swings from positive to negative and back again, and so the current it drives also repeatedly reverses direction, sloshing back and forth within the wire like a rapid tide. AC offers one huge advantage over DC: it presents an elegant solution to the problem of transporting electricity from where it's generated in the power station to where it's needed in industrial complexes or towns.

As soon as you start trying to shunt electrons around a distribution network of metal cables, you hit a fundamental problem. The amount of power delivered by electricity is the product of the current multiplied by the voltage. If you use a large current, the unavoidable electrical resistance of the wires will cause them to warm up and waste the vast majority of the precious energy you've generated. (On the flip side, electrical resistance is the principle that you are deliberately maximizing in the heating element of a coffee maker, toaster, or hair dryer, and if you can get a thin filament hot enough to begin glowing without burning up, then you've cracked the basics of the light bulb, as we've seen.) The only alternative for supplying high power levels is to keep the current low and ramp up the voltage. The problem with this, though, is that high voltages are exceedingly dangerous: acceptable for wires strung high between pylons striding across the countryside, but you certainly wouldn't want them connected to your home. The beauty of AC is that it allows you to easily bump the voltage up and down, using transformers.

A transformer is essentially nothing more than two large coils of wire positioned alongside each other on the same buckle-shaped iron core, so that the magnetic field thrown up by the first coil washes through, the second. Employing the principles of induction discussed earlier, the alternating current flowing through the primary coil creates a rapidly fluctuating electromagnetic field—expanding and collapsing more than a hundred times every second—that in turn induces an alternating current in the secondary coil. Now here's the clever bit. If you wind the secondary coil with more turns than the first, the voltage is stepped up and the current decreases—a transformer is like an electrical currency exchange, interconverting between current and voltage. So you can use transformers to change the voltage in different stages of your distribution network to minimize both the inefficient resistance of high currents and the safety hazards of high voltage.

The beauty of electricity is that you no longer have to build all of your industry on top of windy hills, near fast-flowing rivers, or within easy transport distance of forests or coalfields, as our ancestors had to before the nineteenth century. You only need to place your power generators in these sites, and then zip the electrical energy down wires to wherever it's needed. This is something we've come to take for granted. Just a century ago, all the energy for a household would have to be physically delivered: oil for lamps, charcoal or coal for cooking and heating; Victorian houses needed an outside coal bunker the size of a small room to hold enough fuel to keep warm over the winter. Today, electricity is piped directly throughout the home, supplying energy right to where it's needed—cleanly, silently, and without requiring any storage.

Getting society back on its feet in the immediate aftermath of a cataclysm, DC offers an adequate option for pumping electricity over short distances or storing it in banks of batteries, such as a small-scale local grid of windmills and homes. But if you want to benefit from economies of scale and large centralized power stations as your post-apocalyptic civilization recovers, you'll need to develop an AC distribution network. And in a post-apocalyptic world, where the rebooting society is likely to feel the pinch of much lower energy availability, you will need to make as much use of the heat from fuel as possible. Combined heat and power (CHP) plants address the absurdity of power stations simply discarding vast amounts of heat though their cooling towers, when all the buildings in surrounding towns burn yet more fuel to heat themselves. Sweden and Denmark are leading the world in their use of CHP, first driving turbines to generate electricity but then using the hot steam for other purposes, such as heating buildings in the local area. They are fired by burning natural gas as well as biofuels like wood waste, timber from sustainable forests, or agricultural waste, and can approach 90 percent efficiency for electricity generation and heat production together.

So a familiar sight during the reboot may be animal-drawn carts, or even gasifier-adapted trucks, hauling loads of coppiced lumber and agricultural waste from the surrounding countryside to CHP stations, generating both power and heat for the nearby community and industries to make use of every scrap of the gathered energy. Let's take a look in the next chapter at these transport technologies.

CHAPTER 9

TRANSPORT

A gasoline engine is sheer magic. . . . Just
imagine being able to take a thousand different
bits of metal—and if you fit them all together
in a certain way—and then if you feed them a
little oil and gasoline—and if you press a little
switch—suddenly those bits of metal will all
come to life—and they will purr and hum and
roar—they will make the wheels of a motor car
go whizzing round at fantastic speeds.

ROALD DAHL, *Danny, the Champion of the World* (1975)

MAINTENANCE OF A NATION'S ROAD NETWORK is enormously expensive
and time-consuming, and in the post-apocalyptic world roads would
deteriorate surprisingly quickly, even though the heavy traffic pum-
meling along them will have ceased. In temperate regions, the punishing cycle of freeze-thaw will steadily widen small gaps and cracks, and
seeds blown into crevices will soon grow into stout shrubs and trees,
their roots further crumbling the thin skin of asphalt on the surface.

In fact, our modern asphalt thoroughfares, although beautifully
smooth for bombing along the highway at 70 miles per hour, have a
surface actually less durable than the robust construction of ancient
Roman roads. Many *viae publicae*, crowned with a thick layer of hard

paving stones, were still passable a millennium after the destruction of the civilization that laid them. The same will not be true of our own transport network. Before too long even major highways, the arteries of the old civilization, will become all but impassable. You'll need rugged off-road vehicles even for exploring the dead cities—for the first time SUVs will become necessary to get around urbanized areas.

After the Fall, the solid steel tracks of railways will be far more resilient than roads, but they will eventually succumb to the cancer of rust. Still, over the first few decades long-distance travel over land will probably be easiest along the old train lines, provided you keep them clear of vegetation.

The contraption that underlies much of modern transport is the internal combustion engine: it drives the family car as well as trains and light aircraft. But mechanized vehicles also serve in some of the most crucial roles for supporting society—the tractor, combine harvester, fishing boat, and delivery truck—and you will want to keep these running for as long as possible after the apocalypse. So let's take a look first at how to provide the basic consumables required by mechanized vehicles —fuel and rubber—before exploring what the fallback options might be if society is not able to maintain mechanization and regresses even further after the apocalypse.

KEEPING THE VEHICLES RUNNING

We'll come back in a bit to the slightly different functioning modes of gasoline and diesel engines, but for the moment it is sufficient to understand that they require different liquid fuels. Both gasoline and diesel are liquid mixtures of hydrocarbons—similar molecules to the vegetable oils described in Chapter 5. Gasoline or petrol is a blend of hydrocarbons mostly with backbones 5 to 10 carbon atoms long, whereas diesel is a slightly heavier, more viscous fuel made up of longer compounds of between 10 and 20 carbons. As we saw earlier, substantial reservoirs of these liquid fuels will remain behind after the collapse, in gas stations, depots, and the tanks of abandoned vehicles. But before long the surviving society will need to begin producing its own to sustain mechanized farming and transport.

Today these fuels are made by the processing of crude oil. The methods needed to treat crude oil to yield gasoline and diesel are relatively straightforward, and could be conducted on a small scale by a recovering civilization. Fractional distillation is used to separate out the component liquids, working on the same basic principle as distilling alcohol from water after fermentation. The larger hydrocarbon fractions can be "cracked" to break them down into the more useful smaller-molecule fuels by heating with a catalyst of alumina (such as crushed pumice rock).

So the problem in maintaining a supply of fuels for transport and driving agricultural machinery in the aftermath won't be so much in the difficulty of the chemical processing, but in acquiring crude oil from the bowels of the Earth without sophisticated drilling equipment or offshore rigs. It is possible, though, to make automobile fuel without using oil as the feedstock, and a post-apocalyptic society may learn a lot from the green movement today. Rudolf Diesel himself already noted in the early 1900s that "power can be produced from the heat of the sun, which is always available for agricultural purposes, even when all natural stores of solid and liquid fuels are exhausted."

A viable substitute for gasoline-powered vehicles is ethanol (which we saw in Chapter 5 can be produced by fermentation). Brazil is the world leader in booze-fueled vehicles: every car on its roads runs on an ethanol blend, from 20 percent mixed with gasoline up to 100 percent ethanol-fueled. Even in the United States, many states require that all gasoline contain up to 10 percent alcohol, a blend that can be used without modification to the engine. Indeed, the very first mass-produced car, the Ford Model T, was designed to run on either fossil-fuel gasoline or alcohol, and several distilleries in the US converted crops into car fuel until Prohibition killed the practice.

The problem with large-scale production of ethanol for fueling the transport system of a recovering civilization is sourcing enough refined sugar to feed the fermenting microbes. Crops like sugarcane that underpin Brazil's sustainable biofuels economy cannot be grown outside the tropics. And while sugars are present in all vegetation, making up the strands of cellulose that plants use for structural support, the cellulose is so tough and chemically stable that the vital sugars are locked tightly away and inaccessible. After the Fall, therefore, rather than trying to process such biomass into refined fuel suitable for motor engines, it may be much more feasible to rot it in a biodigester to produce methane gas (see page 74), or simply burn it to fire a boiler in a static power station.

The rumble of diesel engines, on the other hand, will almost certainly still be heard in the post-apocalyptic world. A diesel engine is pretty versatile and can be run on vegetable oil processed into biodiesel by reacting the oil with the simplest alcohol, methanol, under alkaline conditions (by adding lye—either sodium or potassium hydroxide, as we saw in Chapter 5). Methanol, also called wood alcohol, can be produced by dry distillation of lumber (see page 118), but ethanol from fermentation will also do. Any leftover methanol or lye, as well as the undesired side products glycerol and soap, can be cleansed out by dissolving them in water bubbled through the biodiesel, which finally needs to be thoroughly dried by heating to drive off the water before use.

Practically any vegetable oil can be used. Oilseed rape is a good crop as rapeseed yields a great deal of oil per acre (more than other sources such as sunflowers or soybeans), the oil can be easily pressed from the seeds, and the leftover stems make nutritious animal fodder. If need be, animal fats can also be used. Tallow is rendered from scrap meat or carcasses by simmering in water to melt the fat, which then separates and floats, and can be scraped off after cooling. Tallow is processed into biodiesel just like vegetable oils, but the longer hydrocarbons that are present mean that it is liable to congeal in the fuel tank in colder weather.

The issue with these biofuels is that they rely on the transformation of crops into fuel, and keeping even a small car on the road would consume the agricultural output of at least half an acre. Depending on the circumstances of the recovery, it may be that food is scarce for the surviving population. In that case, can vehicles be powered from nonedible sources?

All internal combustion engines actually run on gas (not to be confused with the abbreviation of "gasoline"), rather than liquid fuels. A fine mist of gasoline or diesel is created, which vaporizes before combusting in the cylinder. So another option for keeping mechanized transport going is to deliver combustible gas directly into the engine from a pressurized gas cylinder. This is how modern compressed natural gas (CNG; methane) or liquefied petroleum gas (LPG; a mixture of propane and butane) vehicles are fueled.

A low-tech alternative suitable for the aftermath, when pumping gases into canisters at a pressure of hundreds of atmospheres may prove too challenging, would be to fit vehicles with gas storage bags. Common during the fuel shortages of the First and Second World Wars, these hold coal gas or methane in rubber-sealed fabric balloons, with two to three cubic meters of gas the equivalent of a liter of gasoline.

A slightly less unwieldy option is to build a wood-powered car and generate the fuel gas as you drive.

The key principle is known as gasification. To understand this, light a match and peer in closely. You'll notice that the yellow luminous

A SCOTTISH BUS FUELED BY A GAS BAG DURING THE FIRST WORLD WAR.

flame actually dances clear of the blackening wooden stick, separated by a clear gap. The flame is not in fact predominantly fueled by the matchstick itself, but by combustible gases produced as the complex organic molecules of wood are broken down in the heat, gases that ignite as the hearty flame only when they meet oxygen in the air. This is the same process of pyrolysis we explored in the context of the dry distillation of wood and condensation of the vapors into a variety of useful fluids, but for powering an engine we want to maximize the conversion to flammable "producer" gases and separate the pyrolyzing wood and the flame much farther than in the match. These gases must be prevented from igniting until they can be piped into the engine, where they are finally allowed to mix with oxygen and explode usefully in the cylinders.

During the Second World War, nearly a million gasifier-powered vehicles kept essential civilian transportation running throughout Europe. Germany produced a version of the Volkswagen Beetle with all the wood gasification equipment installed smartly within the body, with a hole in the hood for loading more wood the only giveaway to its extraordinary power source; and the German army even deployed more than fifty Tiger tanks propelled by wood gasifiers in 1944.

A gasifier is essentially an airtight column with a lid on top, and can be constructed from salvaged materials such as a galvanized trash can atop a steel drum, and common plumbing fittings. New wood is piled in at the top, and as it slowly progresses down it is first dried, then pyrolyzed by the contained heat, and partially combusted in the limited oxygen to generate the necessary operating temperature. Most important, a bed of hot charcoal forms at the bottom of the column and reacts with the vapors and gases given off by the pyrolysis to complete their chemical conversion. The final producer gas is then drawn out of the bottom, rich in flammable hydrogen, methane, and carbon monoxide—which is poisonous, so make sure you operate only in a well-ventilated area—along with up to 60 percent inert nitrogen. Cool the producer gas to condense any vapors that could otherwise gunk up the engine, and then feed it into the cylinders.

Around 3 kilograms of wood (depending on its density and dryness) is equivalent to a liter of gasoline, and so the fuel consumption of producer gas cars is measured not in miles per gallon, but miles per kilogram—wartime gasifiers achieved around 1.5 miles per kilogram.

Fuel is not the only consumable needed to keep an automobile rolling. Rubber is required to manufacture the tires that are constantly worn down by driving, as well as the inner tubes that are inflated like doughnut-shaped balloons to cushion the journey.

To be of practical use, the material properties of raw rubber need to be tweaked by vulcanization: it is melted with a sprinkling of sulfur and then poured into a mold to set. In the process the rubber's coiled

VEHICLE POWERED BY WOOD GASIFIER.

molecular chains become interlinked into a tough, resilient mesh by bridges of sulfur. This produces an almost indestructible substance, more elastic than native latex, which doesn't become sticky when warm or brittle when cold.

The trouble with rubber is that once it has been vulcanized, it cannot be simply melted down and re-formed into new products. To provide an adequate supply of tires with crisp treads, as well as providing for all the other uses of rubber, such as valves and tubes, a post-apocalyptic society won't be able to recycle leftovers: it is going to need to find a fresh supply of rubber.

Rubber has been traditionally produced from latex tapped out of the *Hevea* rubber tree, which grows only under humid tropical conditions within a narrow strip around the equator. An alternative source is provided by the stems, branches, and roots of guayule. In contrast to

Hevea, this small shrub is native to the semiarid plateaus of Texas and Mexico. Guayule achieved prominence during the Second World War when the Allies lost 90 percent of their rubber supply with the Japanese invasion of Southeast Asia. The chemistry behind making synthetic rubber will be fiendishly tricky in the early stages of recovery, so once preexisting rubber supplies have deteriorated after the grace period, reestablishing long-distance trade will be one of your top priorities if you don't live near a natural source.

Even if you are able to provide for your fuel and rubber demands, you won't be able to keep vehicles going indefinitely. The components of any remnant machinery will inexorably wear out and deteriorate, and although you will be able to cannibalize spare parts for a certain period, you will inevitably have to begin making your own. Manufacturing replacements for modern engines will demand a high level of metallurgical know-how to blend appropriate alloys and machine tools able to create parts to exacting tolerances—topics that we covered in Chapter 6. And so if the post-apocalyptic civilization does not reattain these capabilities before the last working engine seizes and fails, it will lose mechanization, and the surviving society will regress even further. So in this situation, what backups are available to you to keep the vital functions of transport and agriculture running?

WHAT IF YOU LOSE MECHANIZATION?

If mechanization decays away, the post-apocalyptic society will have to revive animal power. The first beasts in history to be employed as draft animals, hauling carts, wagons, plows, harrows, and seed drills, were oxen—castrated bulls—and they could be co-opted again once the mechanized tractors grind to a halt. Draft horses, such as the Shire horse, descend from animals originally bred to carry fully armored knights across the battlefields of medieval Europe and are faster, stronger, and tire much less readily than oxen. But if you want to replace oxen with horses you'll first have to reinvent the correct harness, a critical accessory that completely eluded the ancient and classical civilizations.

Oxen can be yoked fairly simply with a wooden beam resting across the top of their necks, with staves positioned on either side of the neck to hold it in place, or with a head yoke seated in front of the horns. The body shape of the horse, on the other hand, must be harnessed with an arrangement of straps. The simplest system is known as a throat-and-girth harness, with one strap passing over the top of the shoulders and around the thick neck of the horse, and another underneath the belly, with the load-attachment point positioned over the middle of the back. This style of harness was widely used in antiquity, and served the chariots of the Assyrians, Egyptians, Greeks, and Romans for centuries. But it is actually totally inappropriate for the anatomical structure of the horse and simply doesn't work for hard draft work like pulling a plow. The problem is that the front strap cuts into the horse's jugular vein and windpipe, so that the animal practically throttles itself if it pulls too hard. The solution is to redesign the harness to shift the point through which the animal applies its force.

The collar harness is a well-padded ring of metal or wood that fits snugly around the neck, with the draft-attachment points not behind the neck but lower on either side of the body, so as to evenly distribute the load around the horse's chest and shoulders. This anatomically sound collar—an early application of ergonomic design—was developed in China in the fifth century AD, although it wasn't widely adopted in Europe until the 1100s. It allows the horse to exert its full strength—the animal can deliver three times more tractive force than with the older, inappropriate harness—and horse-drawn plows thus became central to the revolution in medieval agriculture.

MAKESHIFT HORSE-DRAWN TRAPS SUCH AS THIS MAY BECOME COMMON IF MECHANIZATION IS LOST.

The merging of animal traction power and remnant vehicles after the apocalypse will produce some bizarre sights. The working unit of the rear axle and wheels from a defunct car or truck can be salvaged and reappropriated to form the basis of a wooden-sided cart. Even more simple would be to slice a car in half, dump the front end with its inoperable engine, and keep the backseat and rear wheels. The addition of a pair of scaffolding tubes would serve as arms for hitching a donkey or ox for propulsion. Such makeshift traps may become common with the loss of mechanization.

However, reverting to animal power would also require the redirecting of agricultural produce to feeding livestock rather than people. During the peak of animal use for agriculture in Britain and the United States, which surprisingly occurred as late as ca. 1915 (even though mobile steam engines had existed for fifty years and gasoline-powered tractors were already available), a full third of cultivated land was committed to the upkeep of horses.[1]

As well as providing traction power for drawing agricultural tools and transport overland, reclaiming the seas after the apocalypse will be a top priority for reestablishing fishing and trade, and if the capability to maintain mechanization is lost, you'll have to rely on sailing ships.

The most basic mode of a sail makes intuitive sense to anyone who has watched bedsheets drying outside on the clothesline and billowing in the wind. Plant an upright post in the middle of your boat as a mast, and from the top sling a beam horizontally, and at right angles to the length of the hull, for a yard. Dangle a large canvas sheet from the yard, secured with ropes at the bottom, and you have a simple square-sail vessel, which has been independently invented by numerous cultures throughout history. The sail acts to trap the air blowing from behind, and even primitive ships can make good progress running with the wind. But with this rigging you'll never be able to sail closer than about 60 degrees to the wind direction, so you're very much at the mercy of the vagaries of the breeze.

A more sophisticated setup is the fore-and-aft sail. This is not held perpendicularly across the boat, but oriented along the line of the hull, suspended diagonally by a slanted yard or rope attached at one end to the mast. Ships rigged in such a way are far more maneuverable, and

---

1 In fact, there is a recent precedent for just such a technological regression following the crash of mechanization, and an emergency rebooting of animal traction power. Beginning in the early 1960s the agricultural system of Cuba, following Castro's revolution and the adoption of the Caribbean island as a Soviet client state, was transformed by farming machinery and supplies provided by the Soviet Union and East European nations. With the collapse of the Soviet bloc in 1989, however, communist Cuba was abruptly severed from its imported supply of fossil fuels and equipment, and faced a nationwide breakdown of transportation, mechanized agriculture, and the ability to produce fertilizer or pesticides. The nation was forced to rapidly redevelop substantial animal power to replace 40,000 tractors, and an emergency breeding and training program was initiated. In less than a decade Cuba had built up its oxen herds to almost 400,000, as well as a recovering horse population, to keep working its fields.

can also tack and beat much closer to the wind—modern yachts can cut as little as 20 degrees into the wind—than a square-rigged vessel, although most large ships set a combination of both kinds of sail. Fore-and-aft rigging dates back to Roman navigation of the Mediterranean, but really came into its own during the Age of Discovery, hauling the great European exploration ships, led by the Portuguese and Spanish, across the oceans of the world to encounter distant new lands and establish long-range trade routes.

When you present a fore-and-aft sail obliquely to the wind, a whole new effect comes into play. The wind filling the sail causes it to bulge outward and behave like an airfoil—the airflow rushing over the curved surface is deflected and creates a region of low pressure in front of the sail. Rather than being blown through the water with the wind drag created by a square sail, the fore-and-aft sail is sucked forward by this aerodynamic lift force. So without fully understanding the physics involved, in 1522 Ferdinand Magellan's expedition became the first to circumnavigate the Earth using the same aerodynamics that lies behind the aircraft wing and the reaction turbine.

Using fore-and-aft rigging to catch the wind blowing across the boat, you have now created a stability problem, though, and your vessel is at risk of being rolled right over and capsizing. The solution is to load ballast low down in the ship to keep it self-righting, and to fit a keel below the hull, often shaped like an upside-down shark's fin, to resist the tipping force of the sails. But if you are able to control these competing forces, and carefully adjust the rigging to trim the fore-and-aft sails into the optimal curve, the astounding consequence of the physics underlying their airfoil effect is that they can actually sail faster than the wind blowing them.

If you can't salvage any serviceable hulls, you will need to build your own. Traditional shipwrighting involves fixing planks lengthwise to a frame and rendering the seams watertight by stuffing them with plant fibers caulked with pine pitch; or if you can scavenge or smelt enough wrought iron or steel plates, you can rivet them together. Sails are essentially large sheets of fabric, an application of the same weaving technology we encountered in Chapter 4. When creating a sail, use a plain weave, and be aware that any fabric is strongest when pulled along the direction of the weft, as these threads are already straighter than the warp, and the material is readily distorted and potentially damaged if stretched along a diagonal (try this with a small section of your shirt now). Likewise, the ropes rigging everything together are produced by spinning fibers into yarn, with the yarn then twisted into strands and strands into ropes—and, if need be, ropes into cables. The pulleys and tackle blocks needed for controlling the sails are identical to those used for hoisting heavy loads on scaffolding or cranes on a building site.

Hopefully, before too long the recovering civilization will once again begin mastering metalworking and machine tools. One mode of mechanically simple transportation for personal conveyance in a post-apocalyptic world without working motors would be the bicycle. The heart of the pedal-powered bike is the crank that converts the up-down stroke of your legs into a rotary motion applicable to wheels. But there is still a major engineering problem to solve: you can't directly couple this cycling action to the wheel, with pedals fixed to the axle like a child's trike, because for any meaningful velocity you'd need to pump your legs like one possessed.

The simplest approach is therefore to fit a large front wheel, so that even with a modest rotation the enormous circumference confers a decent speed; this was the idea behind the ludicrous-looking penny-farthing with its four-foot wheel. A far better solution, which seems so obvious to us now but wasn't conceived by a bicycle manufacturer until 1885, is to use gears, an ancient mechanical system, linked with a chain. Two sprockets of different sizes allow the driven wheel to rotate much faster than the pedal crank and are mechanically coupled by a roller chain (itself very similar to a design sketched by Leonardo da Vinci in the sixteenth century). Another key working principle is that the front upright, linking the hub and handlebars, should be tilted backward slightly so that the leading wheel naturally steers into any sideways topple to bestow inherent stability on the bicycle.[2]

REINVENTING POWERED TRANSPORT

At some point, the recovering civilization will reattain the sophistication in metallurgy and engineering necessary to contemplate building engines. If the post-apocalyptic society has regressed to the stage of relying on draft animals and sails, how could it now reinvent the internal combustion engine without reference to any surviving examples? What's the anatomy of the heart that throbs under the hood of our vehicles?

The internal combustion engine is a great example of how any complex machinery is no more than an assemblage of basic mechanical components, all with very different heritages, and arranged in a novel configuration to solve the particular problem at hand. If you could peel back the metallic skin and dissect the family car as an organism, you'd find myriad different sub-mechanisms interacting with one another like the various organs and tissues in the human body.

So what are the key principles behind the functioning of the automobile, and how could you design one from scratch if you needed to?

We've already looked in Chapter 8 at the operating principle of an *external* combustion engine: the steam engine works by burning fuel to heat a boiler and force steam into a cylinder. A far more efficient use of

---

2 Contrary to popular belief, a bicycle's stability has little to do with the gyroscopic effect of its spinning wheels, especially at slow speeds.

the chemical energy locked in the fuel is to cut out the middleman and use the pressure of hot gas produced by the burning *itself* to drive machinery. If a tiny amount of fuel is introduced into a confined space before being ignited, the explosive expansion of the resultant hot gases can be made to shunt out a piston and so perform work for you. Do this several times every second and you've got a regular, reliable delivery of power. To reset the cylinder for another burst, open a hole and push the piston back to squirt out the exhaust gases like the plunger of a syringe, then draw it out again to suck oxygen-laden air laced with fresh fuel through a second valve. Begin compressing this mixture to get it dense and hot before igniting it again. This four-stroke cycle is the rapidly beating heart of most internal combustion engines on the planet.

You've got two options for triggering the combustion of the fuel once it's in the cylinder, and this marks the difference between modern gasoline and diesel engines. Volatile fluids like ethanol (or gasoline) can be vaporized by mixing them with air in the carburetor before being introduced into the cylinder and ignited with an electrical spark plug. Mixtures of heavier hydrocarbon molecules like diesel can be sprayed into the cylinder as a fine mist at the end of the compression stroke, to vaporize and ignite spontaneously as a consequence of the temperature surge from extreme pressurization of the air. (Anyone feeling the nozzle of an air pump after filling their tires will have noticed how warm it can get from the air compression.) Or you could, as we saw at the beginning of the chapter, fuel your engine with gas piped directly into the cylinders.

For powering a vehicle, though, the challenge now is to convert the reciprocating back-and-forth movement of the pistons into a smooth rotation that can be used to spin wheels or a propeller. The device that performs this crucial translation of motion is the crank, as we saw with the bicycle. The crank is often used in machinery, with a pivoted connecting rod linking the reciprocating component and rotating shaft (on a bicycle, it is your leg that comprises the connecting rod coupled to the pedal crank). The earliest known appearance of such a crucial mechanism was installed on a third-century-AD Roman waterwheel, where it converted river-powered rotation into the drawing back and forth of long wood saws.

Modern engines, which team up the power of multiple firing pistons, employ a slight modification known as a crankshaft, which has a series of handle-like kinks spaced along its length to allow a whole row of pistons to drive rotation of the same spindle. Even with several cylinders firing in a staggered sequence, the explosive impulses turning the shaft are still jerky, and a way to even out the rotation is needed. This time the solution is provided by ancient pottery technology. A flywheel is stuck on the end of the crankshaft and functions in exactly the same way as the heavy stone disk on a potter's wheel, storing rotational momentum and smoothing the spin.

Another ancient mechanical component is needed to coordinate the opening and closing of the valves that admit fuel and expel exhaust gases from the cylinder during the power cycle. The cam has an elongated, off-center shape, so that as it turns on a shaft it can be used to rhythmically lift up a lever or push away a "follower" rod. In our history cams were employed in trip-hammers, where the power of a waterwheel was used to repeatedly raise a heavy hammer and then drop it for a blow as the cam catch passed and released it. Cams were known to the ancient Greeks and reappeared in medieval machinery in the fourteenth century. In the modern combustion engine a set of cams, driven from the main crankshaft, allow the operation of the inlet and outlet valves to be perfectly timed with the piston cycle.

If you're intending to use your engine to power land vehicles, rather than to simply spin the propeller on a boat, there are a few more technical challenges you need to solve. With the core design of the engine

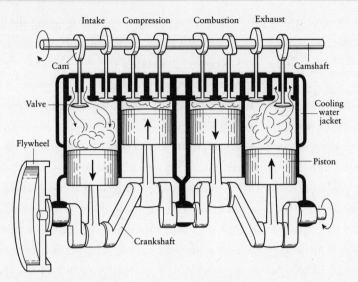

THE FOUR-STROKE INTERNAL COMBUSTION ENGINE, MADE UP OF CYLINDERS
AND PISTONS, A CRANKSHAFT TO DELIVER THE POWER TO THE FLYWHEEL,
AND A CAMSHAFT TO COORDINATE THE OPENING AND CLOSING OF VALVES.

settled, the next mechanical problem is delivering that driving force to
the wheels. One of the most intuitively comprehensible parts in an
automobile power plant is the transmission: in essence it is no more
than a box that allows you to change which pairs of gears are meshed
together, operating on the same basic principle as gear chains dating
back to the third century BC. The internal combustion engine turns
(or revs) at a very high rate, and so low gear ratios, whereby the drive
shaft gear is larger than the gear it is engaged with on the engine shaft,
are used to trade spin speed for turning force. This greater torque is
needed particularly for accelerating or climbing hills.

A related piece of equipment facilitating gear changes is the clutch.
In many automobiles this assembly bears the engine power through a
roughened disk in firm contact with the flywheel—ironically, it is friction that allows the smooth operation of the motor. The disk and fly-

wheel can then be pulled apart to disconnect the engine from the drive shaft. Similar systems were used in early woodworking tools such as the lathe to allow the mechanism to be disengaged from its power source.

The first cars cut and pasted bicycle technology, and drove the rear axle with a chain and sprocket. A more efficient method to transfer the engine power is a spinning drive shaft, but it must be afforded a degree of flexibility to avoid it snapping with the jolts of driving. How, then, do you allow a rigid rod to bend or flex in any direction while still conferring power? The solution lies in positioning two universal joints along its length. Each of these is composed of a pair of connected hinges, a design that was first depicted in 1545.

Once you've got your vehicle tearing along, the next pressing issue is to devise a means by which to conveniently steer the wheels from the driving seat. The earliest cars used a tiller, borrowed directly from marine technology for controlling a boat's rudder. But with a little more thought a far better solution was found, this time co-opting technology originating in ancient water clocks dating back to around 270 BC. The rack and pinion is a mechanism formed by combining a pinion gear and a long bar cut with corresponding teeth. The steering wheel in the cab is linked by a shaft to turn the pinion, which displaces the rack sideways left or right to angle the front wheels.

One final engineering issue arises when you have two wheels fixed on the same axle. As the car rounds a corner, the outer wheel needs to turn slightly faster than the inner, and if they are rotationally locked together both can end up slipping or dragging, making it difficult to steer and damaging the tires. A system known as the differential, an assemblage of no more than four gears, allows both wheels to be driven by the engine while turning at different rates. This ingenious device was applied in European mechanisms from 1720, and possibly dates as far back as 1000 BC in China.

So if you peel back the skin of a brand-new sports car, something you might consider to represent the peak of modern technology, you'll find a mishmash of components co-opted from mechanisms stretching far back through time: potter's wheels, Roman sawmills, trip-hammers, wood lathes, and water clocks.

The internal combustion engine is a miraculous contraption, able to transform the chemical energy latent in fuel into smooth motion, and it underlies much of transport today (alongside the jet engine for fast aircraft and the steam turbine of large ships). We've looked at ways to produce your own gas or liquid fuels to feed these engines, and a full fuel tank offers such a fabulously dense reservoir of energy for traveling great distances before needing resupply that combustion will surely once again play a role in long-range overland and marine transport within a post-apocalyptic society well matured in its recovery. The problem, though, is that without easily accessible crude oil the civilization after ours may well be limited in its fuel sources: the proliferation of motor vehicles from the 1920s on was enabled by the cheap availability of gasoline from oil refineries. So, what might be an alternative developmental pathway for establishing transport infrastructure in a society rebuilding from scratch?

Rather than cultivating crops and picking out only a portion of the plant to be pressed to biodiesel or fermented to ethanol, it may be better to simply burn the whole harvest. Heating boilers to drive steam turbines and generate electricity makes a far more efficient use of the total sunlight energy captured by fast-growing biomass crops like switchgrass or miscanthus, or coppiced woodland. The electricity supply generated sustainably from biofuels as well as wind power and hydropower can be shunted down overhead wires to power trains and trams along fixed routes, or to recharge batteries for smaller vehicles. An electric car can travel farther on an acre of crop than an internal combustion engine filled with biofuel from the same harvest, and, what's more, a boiler driving a steam turbine can be fired with far lower-grade plant matter than is required for biofuel synthesis. And if you generate that electricity in a combined heat and power (CHP) plant, you can use the waste heat to warm buildings in the vicinity. An energy-restricted post-apocalyptic society will need to use coordinated thinking to maximize the efficiency of its fuel consumption, and it seems likely that the urban transportation of a post-apocalyptic civilization will be predominantly electric.

In fact, electric vehicles have been common once before. In the early years of the twentieth century, there were three fundamentally different automobile technologies battling for supremacy, and electric cars held their own against competition from steam- and gasoline-powered alternatives, as they are mechanically much simpler and more reliable, as well as quiet and smokeless. In Chicago they even dominated the automobile market. At the peak of production of electric vehicles in 1912, 30,000 glided silently along the streets of the USA, and another 4,000 throughout Europe; in 1918 a fifth of Berlin's motor taxis were electric.

The drawback of electric cars with their own onboard batteries (rather than trains or trolleys taking a continuous feed from a power line over the track) is that even a large, heavy set cannot store a great deal of energy, and once depleted the battery takes a long time to recharge. The maximum range of these early electric vehicles was around a hundred miles,[3] but this is farther than a horse and in an urban setting is more than adequate. The solution is, rather than waiting for the battery to be recharged, you can simply pull into a station for a quick battery pack exchange: Manhattan successfully operated a fleet of electric cabs in 1900, with a central station that rapidly swapped depleted batteries for a fresh tray.

---

3 Ironically, about 100 miles is still the maximum range for modern electric cars: technological improvements in battery storage and electric motors have been perfectly offset by an increase in car size and weight, and drivers of electric vehicles suffer from "charge anxiety."

So with a combination of biofuel-fed internal combustion engines and electric vehicles, an advancing post-apocalyptic society will be able to provide for its transportation requirements even without the abundant oil that we benefited from in our development. Now it's time to turn from the transport of people and materials to the conveyance of ideas: in the next chapter we'll explore communication technologies.

CHAPTER 10

COMMUNICATION

I met a traveler from an antique land
Who said: "Two vast and trunkless legs of stone
Stand in the desert. Near them, on the sand,
Half sunk, a shattered visage lies, whose frown,
And wrinkled lip, and sneer of cold command,
Tell that its sculptor well those passions read
Which yet survive, stamped on these lifeless things,
The hand that mocked them and the heart that fed:
And on the pedestal these words appear:
'My name is Ozymandias, king of kings:
Look on my works, ye Mighty, and despair!'
Nothing beside remains. Round the decay
Of that colossal wreck, boundless and bare,
The lone and level sands stretch far away."

PERCY BYSSHE SHELLEY, "Ozymandias" (1818)

TODAY, WITH THE INTERNET, ubiquitous wireless networks, and hand-held smartphones, communication with one another anywhere in the world is effortless and instantaneous. We keep in touch via e-mail and Twitter, websites disseminate news and information, and we can access the wealth of human knowledge from the palm of our hand. But in a post-apocalyptic world you'll need to return to more traditional communication technologies.

WRITING

Before the invention of writing, knowledge circulated among the minds of the living, conveyed only by the spoken word. Yet there is only so much data that can be stored in oral history, and the danger is that when people die ideas are lost forever. But once committed to a physical medium, thoughts can be stored faithfully, referred back to years later, and built up over time. A culture that has developed writing can accumulate far more knowledge than could ever be cached in the collective memories of its populace.

Writing is one of the fundamental enabling technologies of civilization. It involves the conceptual leap of transforming spoken words into sequences of drawn shapes: either arbitrary letters representing the individual sounds of the language (such as the phonemes of English) or characters symbolizing particular objects or concepts (like the morphemes of Chinese). At the basic level, it allows you to permanently record the agreed terms of trade, a land lease, or a code of laws. But it is the accumulation of *knowledge* that allows a society to grow culturally, scientifically, and technologically.

In the modern world we've come to take for granted such staples of civilization as pen and paper, and realize how vital they are only when we can't simply reach for the back of an envelope to jot down a shopping list, or when we bemoan the confounding disappearance of the ballpoint we put down only two minutes ago. While plentiful paper will be left behind by our civilization, it is a particularly perishable material and will readily burn with the wildfires tearing through deserted cities or molder away with humidity and floods. How can you easily mass-produce paper for yourself, and leapfrog over the time-consuming production of other materials, such as papyrus and parchment, used historically?

Paper was invented by the Chinese sometime around 100 AD,

although it took more than a millennium to diffuse across to Europe. Paper made from tree pulp, though, is a surprisingly modern innovation. Until the late nineteenth century, paper was mainly manufactured from linen fragments, recycling tattered rags. Linen is a fabric made from fibers of the flax plant (see Chapter 4), and any fibrous plants can in principle be converted into paper: hemp, nettles, rushes or other coarse grasses. But as demand grew, spurred on, as we'll see, by the plethora of books and newspapers churned out of the printing presses, other suitable fibers were intently sought. Wood is a fabulous source of good-quality papermaking fibers, but how do you disassemble a thick, solid tree trunk into a fine soupy mush of soft, short strands without breaking your back in the process?

The fibers that make paper so light yet strong are composed of cellulose. Chemically this is a long-chain compound used by all plants as the main structural molecule between their cells, and in particular in their stem and side shoots; it is the pithy strands of cellulose that get stuck between your teeth when you munch on celery. In the stout trunks of trees and shrubs, however, the cellulose fibers are reinforced with another structural molecule called lignin, which locks the cellulose strands together to make wood. This provides the tree with the ideal structural material for a strong, load-bearing central column and wide-spreading branches to splay its leaves out before the Sun, but it makes the cellulose fibers lamentably inaccessible to us.

Traditionally, plant fibers were separated by crushing the stems and then retting—soaking them for several weeks in stagnant water to allow microorganisms to begin decomposing the structure—and then violently pounding the softened stalks to liberate the cellulose fibers by brute force. The good news is that you can save yourself a great deal of time and effort and leapfrog straight to a much more effective scheme.

The links that bind together cellulose and lignin in trees are vulnerable to the chemical severing process known as hydrolysis. This is the same molecular operation that is employed in saponification during soap making, and we achieve it with exactly the same means: by rallying alkalis to the cause. The best parts of the tree or plant to use are the stem or trunk and branches—the roots and leaves don't contain much of the cellulose fiber required. Chop the material into small pieces to expose as much surface area to the action of the solution as possible, then bathe it in a vat of boiling alkaline solution for several hours. This breaks the chemical bonds holding together the polymers, causing the plant structure to soften and fall apart. The caustic solution attacks both cellulose and lignin, but the hydrolysis of lignin is faster, allowing you to liberate the precious papermaking fibers without damage while the lignin degrades and dissolves. Short white fibers of cellulose will float to the top of the murky brown, lignin-stained broth.

Any of the alkalis we covered in Chapter 5—potash, soda, lime—work, though the preferred option through much of history has been to use slaked lime (calcium hydroxide), as it can be generated in bulk by cooking limestone, while potash is fairly labor-intensive to produce by soaking timber ashes. But once you've cracked the artificial synthesis of soda (we'll come to this in Chapter 11), the best option by far for chemical pulping is to use caustic soda (sodium hydroxide), which powerfully promotes hydrolysis. You generate this directly in the pulping vat by mixing together slaked lime and soda.

Collect the recovered cellulose fibers in a sieve and then rinse several times until they run clear of the mucky lignin color. To lighten the shade of the finished paper to a clean white, you can also soak the pulp in bleach at this point. Calcium hypochlorite or sodium hypochlorite are both effective bleaching agents, and can be created by reacting chlorine gas (produced electrolytically from seawater—see page 232) with slaked lime or caustic soda, respectively. The chemistry behind this bleaching effect is oxidation: bonds in the colored compounds are broken to destroy the molecule or convert it to an uncolored form. Bleaching is critical not only to papermaking, but also to textile production, so it will likely be a key driving force for expanding the chemical industry during a reboot.

Pour a dollop of this sloppy cellulose soup across a fine wire mesh or cloth screen, bounded on the sides by a frame, so that the fibers form a higgledly-piggledy mat as the water drains out. You then press it to squeeze out the remaining water and to ensure flat, smooth sheets of paper, and leave to dry.

You'll find small-scale paper production much easier if you're able to scavenge a few items from the fallen civilization. A wood chipper or even a large food processor, powered from a generator, will make lighter work of the chewing up of plant matter into a thick vegetative soup: but you can also let windmills or watermills provide the mechanical brawn needed for driving trip-hammers to pound the material.

However, creating clean, smooth paper is only half of the solution to being able to use writing for communication and recording permanent stores of knowledge. The other critical task, once all of the remnant ballpoints have dried up or disappeared, is to make your own reliable ink with which to form the written word.

In principle, anything that irritatingly stains your cotton shirt if you accidentally splash yourself can also be used as a makeshift ink. You can take a handful of intensely colored ripe berries, for example, and crush them to release their juice, strain to remove the mashed fruit pulp, and dissolve in some salt to serve as a preservative. The major problem with most plant extract inks, though, is their impermanence. To preserve your words and the recovering society's newly accumulated knowledge indefinitely, you really want an ink that won't readily wash off the page or fade in sunlight. The solution that emerged in medieval Europe is known as iron gall ink. In fact, the history of Western civilization itself was written in iron gall ink. Leonardo da Vinci wrote his notebooks with it. Bach composed his concertos and suites with it. Van Gogh and Rembrandt sketched with it. The Constitution of the United

States of America was committed to posterity with it. And a formulation very similar to the original iron gall ink is still in widespread use in Britain today: registrar's ink, required to be used for legal documents such as birth, death, and marriage certificates, uses exactly the same medieval chemistry.

As the name reveals, the recipe for iron gall ink contains two main ingredients: an iron compound and an extract from plant galls. Galls appear on the branches of trees such as oak, and are formed when parasitic wasps lay their eggs in the leaf bud and irritate the tree into forming a growth around it. They are rich in gallic and tannic acids, which react with iron sulfate—created by dissolving iron in sulfuric acid. Iron gall ink is practically colorless when first mixed, and so it's difficult to see where you're writing unless another plant dye is also included. But with exposure to the air, the iron component oxidizes to turn the dry ink a deep, enduring black.

A rudimentary pen can also be made in the time-honored fashion. Soak a bird's feather (goose or duck was preferred historically) in hot water and pull out the material within the shaft. Bring the tip into a sharp point by cutting into each side, and then undercut the bottom face into a gentle curve to create the classic shape of a writing nib. Slitting backward slightly into the pointed tip will allow the nib to hold a tiny reservoir of ink as you write, between replenishing dunks into the inkwell.

PRINTING

If writing is the critical development to enable the permanent storage and accumulation of ideas, then the printing press is the machine for the rapid replication and extensive dispersal of human thought. Today, the developed world boasts near-universal literacy, and an estimated

45 trillion pages are printed every day: books, newspapers, magazines, and pamphlets.

Without printing, if you wanted a document reproduced it would take a dedicated team of scribes arduously copying it by hand for weeks. Hence only the powerful and well resourced would be able to afford the project, which also means only approved or endorsed texts are propagated. But with the dissolution of such a choke point, thanks to the printing press, knowledge becomes democratized. Not only does learning become available to everyone in society, but anyone can rapidly disseminate their own ideas, from new scientific theories to radical political ideologies, encouraging debate and promoting change.

The basic principle of printing is that a page of writing is re-created as rows of types—cuboidal blocks, each with a letter embossed on the top face—arranged within a rectangular frame. The type is inked and then pressed onto a sheet. Once the frame has been typeset, the same page of text can be replicated again and again exceedingly quickly, and when done, the letters are simply rearranged into another page of text. Even a rudimentary printing press can reproduce a document hundreds of times faster than a scribe.

There are three major challenges that you'll need to solve for a post-apocalyptic resurrection of the movable-type printing press, which Johannes Gutenberg invented in fifteenth-century Germany.[1] You'll need to find a way to easily produce large numbers of precisely

---

[1] So why was it that the Chinese developed paper a good millennium before it became common in Europe, and also published texts using wooden block printing, but never made the step to movable-type printing taken by Gutenberg? The historical reasons probably come down to a fundamental difference between the nature of European writing and Eastern scripts. Western writing is made up of a small set of letters rearranged into combinations to spell out the sounds of different words, whereas written Chinese is composed of a vastly greater number of complex compound characters, each symbolizing a particular object or concept. The simple rearrangement of Western letters lends itself to movable-type printing.

sized types. You'll also need to devise a mechanism to provide an even but firm pressure to apply the print to the page. And third, you'll need to invent a new kind of ink that doesn't flow freely from a pen nib, but sticks well to intricate metallic detail.

The first issue you're faced with is what material you use to make the types. Wood can be carved easily, but this would necessitate the diligent work of a skilled craftsman to hand-make each and every piece of type individually—around eighty letters (both lowercase and capitals), numbers, punctuation marks, and other common symbols—and then produce multiple, identical copies of each. And all that hard work for just a single set of type, in only one font size and one style.

So in order to mass-produce printed books, you must first mass-produce the tools for printing. This can be achieved by type casting: founding identical letter blocks with molten metal. The solution for creating types with straight, smooth sides and perfect right-angle edges that slot perfectly alongside each other in rows, Gutenberg realized, is to cast the types in a metal mold with a sharp cuboidal interior void. Cleverly, the crisp shape of a particular letter can be formed on the end face of the block by positioning a swappable matrix at the bottom of the mold. These matrices can be made from a soft metal such as copper, and the precise indent of a letter hammered into each of them very simply with a hard steel punch. Now all you have to do is engrave each letter, number, or symbol just once onto different punches, and you can effortlessly churn out countless pieces of identical type.

There is one final problem, though, thrown up by the nature of the letters in Western script, which is the large variability in their girth: the svelte "i" or slender "l," compared with the rotund "O" or broad-shouldered "W." To be read easily the letters should huddle closely together without gaping spaces around the skinnier letters and numbers. The upshot is that you need to be able to cast cuboidal types that are all exactly the same height, so that they print uniformly on the page, but each with a different width.

The solution is the final spark of Gutenberg's inspiration in devising an elegant system for mass-producing the building blocks of printing. Create the mold in mirror-image halves: two L-shaped parts facing each other to create a cuboid space between them. The walls of this cavity can be simply slid toward or away from each other to smoothly adjust the width of the mold, without changing the depth or height (try it with your thumbs and index fingers now to see how this ingenious system works). Casting a perfectly formed type is now as simple as placing the relevant stamped matrix at the bottom of the mold, setting the width, pouring in the molten metal, and then ejecting the finished piece when it has set by parting the L-shaped halves again.

MOLD FOR TYPE CASTING. THE MATRIX, BEARING THE IMPRINT OF THE
LETTER STAMPED INTO IT, LIES AT THE BOTTOM OF THE CENTRAL CAVITY.

After a page of text is typeset, the type face is inked and transferred as an intricately detailed impression onto a blank sheet. There are a range of mechanical devices that enable the application of enhanced force, such as the simple lever or a pulley system, and both have been used throughout history for squeezing out excess moisture in paper production. Gutenberg grew up in a wine-growing region of Germany, and so co-opted another ancient device for his groundbreaking invention. The screw press is a Roman technology dating back to the first century AD, used extensively for juicing grapes or extracting oil from olives. It also provides the ideal compact mechanism for applying a firm but even pressure onto two plates, squeezing the inked type onto the page. This key component of printing survives to this day in our collective name for the newspapers, and by extension the journalists who report for them: "the press."[2]

The availability of paper is not a prerequisite for the printing press, as the technique also works on parchment made from calfskin (but not on brittle papyrus sheets). But without mass-produced paper, printed books could never be produced cheaply enough for the general population, and so their social-revolutionary potential would remain unrealized. If the book you hold in your hands now had been published in the same typographic format as Gutenberg's first Bible on parchment pages, each copy would require the complete hides of around 48 calves.

Successful printing does rely on a suitable ink, however. The free-flowing water-based inks developed for handwriting, like iron gall ink,

---

2 If you anticipate that you'll want to run further impressions of the same body of text in the future, such as for subsequent print runs of an important treatise, you can save yourself the hassle of having to typeset thousands of individual letters all over again by saving the page configuration. The types themselves are too valuable to be left arranged in the frame, but you can take an impression of the text layout in plaster and then use this as a mold for casting a metal plate of the whole page. This is the original meaning of the word "stereotype." The nickname for a stereotype plate is "cliché," apparently after the sound made during the casting—and so to use a cliché is to rehash a block of commonly printed text.

are totally inappropriate for printing. To print crisp lettering you need a viscous ink that will stick readily to the metal features of the detailed type and then transfer cleanly to the paper without smearing, running, or blurring. Gutenberg solved this particular challenge by borrowing from a fashion that was only just starting among Renaissance artists: the use of oil paints.

Both the ancient Egyptians and the Chinese developed a black ink based on soot at around the same time, roughly four and a half thousand years ago. The soot's tiny carbon particles serve as a perfectly dark pigment when mixed with water and a thickener, such as tree gum or gelatin (animal glue—see Chapter 5). This is the composition of India ink, which, despite its name, was actually first developed in China and traded with India, and is still popular with artists today. Indeed, a suspension of carbon-black pigment particles also forms the basis of photocopier and laser-printer toner. Soot particles can be trapped from the smoky flame of burning oils—a substance known as lampblack—as well as by charring organic materials such as wood, bone, or tar.

While carbon-black pigments have a long heritage, the glue- or gum-thickened India ink is not suitable for a printing press: you need an ink with very different viscosity (runniness) and drying behavior. And this is where Gutenberg borrowed from the very beginnings of Renaissance oil painting. Lampblack mixed into linseed or walnut oil dries well and sticks to metal type far better than a runny, water-based ink. (Linseed oil does need to be processed before being used, though: boil it and remove the thick, gluey mucilage that separates on top.) You can control the ink's crucial viscosity with two other ingredients, turpentine and resin. Turpentine is a solvent used for thinning oil-based paints, and is produced by distillation of resin tapped from pines or other conifer trees (see page 119). The hard, solidified resin left behind after the volatile compounds have been driven off during distillation, on the other hand, will thicken the solution. By tweaking the balance of these two contrary constituents you can perfect the viscosity of the ink, and you can control its drying behavior by varying the proportion of walnut and linseed oils.

So printing can rapidly replicate knowledge through your recovering civilization, and long-distance communication can be achieved by sending written messages. But how might you use electricity to communicate over great ranges without having to go through all the bother of physically transporting the message?

ELECTRICAL COMMUNICATIONS

Electricity is wonderful stuff: it will shoot virtually instantaneously along a wire laid for it and produce a noticeable effect far away from the operating switch—illuminating a light bulb in another room, for example. But to communicate between buildings, cities, or even continents, you can't simply extend a light bulb–powering circuit and flash messages to one another. Energy-sapping resistance is your enemy here, as there won't be enough voltage to power a bulb over any meaningful distance. A good electromagnet, built as we saw in Chapter 8, will generate an appreciable magnetic field from even a feeble current, though. Position a lightly balanced metal lever over the end and you can use it as an exquisitely sensitive switch, pulled closed to sound a buzzer whenever the electromagnet is energized. A relay-controlled buzzer at both ends of a long telegraph wire will allow remote operators to hear whenever either is sending current.

Messages can be sent one letter at a time by representing them as combinations of short or long squirts of current—dots and dashes. All you need to do first is agree with the guy at the other end of the telegraph cable how you will represent each letter of the alphabet, and then send your first post-apocalyptic e-mail down the wire. Exactly how you organize this doesn't really matter, but with a little bit of forethought on how to ensure that the coding system is both rapid and reliable, you'll probably reinvent something similar to Morse code. In this system, the most commonly used letters in the English alphabet are represented by the simplest forms: E is a single dot, T is a single dash, A is dot-dash, and I is dot-dot.

Regularly spaced relay stations will boost the current along the next section of wire and so allow globe-spanning telegraphic communications. But the laying and maintenance of wires draped across the continents and ocean floors is difficult. So is there a better way? Can you communicate using electricity, but without the bothersome wires needed to carry current?

Let's look more closely at the yin-yang relationship of electricity and magnetism. If a changing electric field can generate a magnetic field, and a changing magnetic field can in turn induce an electric field, then you ought to be able to create a ripple of mutually supporting energies. Indeed, such electromagnetic waves propagate even through a perfect vacuum with no matter present to carry the disturbance (unlike a sound or water wave): electricity and magnetism combine to travel like ghosts through the universe.

The golden sunlight streaming through my window is itself nothing more than a welding of electric and magnetic fields. Indeed, everything from X-ray machines, ultraviolet tanning beds, infrared night-vision cameras, and microwave ovens to radar, radio and TV broadcasts, and the ultimate expression of modern life, this free Wi-Fi hotspot I've hopped onto with my laptop, are all based on varying forms of light. The electromagnetic spectrum is a broad swath of waves with different frequencies of vibration of the coupled electric and magnetic fields, stretching from dangerously energetic gamma radiation to long-wave radio, but all propagating at the speed of light.

But it is radio waves that interest us here. Not only are they relatively simple to make and catch, but they can also be imprinted with information to carry over vast distances. It is this radio transmitter and receiver technology that you'd ideally want to recover as a means for long-range communication during the reboot.

Let's start with the slightly easier task of building a radio receiver. Dangle a long piece of wire from a tree, with the bottom end stripped of any insulation and buried in the earth to ground it. This is your aerial, and the rapidly fluctuating electromagnetic fields of any passing radio waves will drive electrons in the metal to slosh up and down the wire—an induced alternating current. But in order to drive a pair of earphones to hear anything, you need some way of keeping either the negative or positive parts of the wave and discarding the other half.

Any material that allows electricity to flow through it in only one direction, blocking the reverse tide, will accomplish this, "rectifying" an alternating current into a series of pulses of direct current. Luckily, many different crystals turn out to exhibit this marvelously useful property. Iron disulfide, also known as fool's gold for its deceptive appearance, works well and is easy to spot. Another mineral, galena (lead sulfide), is also used commonly in crystal radio sets. Galena is the main ore of lead, found in large deposits all over the world, and has been mined throughout history to produce plumbing pipes, church roofs, musket shot, and rechargeable lead-acid batteries.

Connect the crystal into your aerial-earphone circuit by placing it in a metal holder, and make a second contact with it using thin wire, known as a cat's whisker. Rectification happens at the connection between the crystal and the point of contact, but the effect is elusive, and finding a sweet spot by trial and error requires a lot of patience. Nonetheless, even in the absence of any human broadcasts, this rudimentary setup may allow you to pick up the radio emissions from natural sources like lightning storms. In fact, a rudimentary radio transmitter—the spark gap generator—works by creating a rapid series of artificial lightning discharges.

Spark gap generators leave a small gap in a high-voltage electric circuit so that a spark repeatedly leaps across it. Each spark releases a surge of electrons along the aerial and the emission of a brief burst of radio waves. If the transmitter circuit sparks thousands of times a second, releasing a rapid train of radio pulses, a buzzing tone will be heard in the earphones of receiver sets. Insert a switch on the low-voltage side of the transformer that powers the spark gap to control when the circuit is energized and transmits radio waves, and again encode your message in dots and dashes.

Ideally, though, you want to be able to transmit sound over the radio waves, allowing conversations between individual radio operators or the broadcast of news to a widely spread audience. Morse code involves crudely turning the radio waves completely on or off, but conveying sound requires a more refined manipulation, known as modulation of the carrier wave. The simplest scheme is called amplitude modulation (AM), whereby the intensity of the carrier wave is varied more smoothly between these two extremes: the gentle contours of the sound wave are imprinted on top of the frenetic fluctuations of the radio wave. Thankfully, the cat's-whisker crystal detector also works admirably to "demodulate" the signal in the receiver. The one-way-street behavior of the crystal junction coupled with the smoothing effect of a capacitor strips away the high-frequency carrier wave, leaving behind the broadcaster's voice or music.

Unless you have only a single high-powered transmitter nearby, the signal you hear with this bare-basics radio receiver will be a confused mash-up of stations: the aerial picks up a variety of transmissions on different frequencies of carrier wave and passes them all on to your earphones. Adding a few extra components to your electronic machines will allow you to tune these radio sets. Tuning makes a radio transmitter more efficient by packing the broadcast energy into a narrow span of radio frequencies, and a tuned receiver plucks only the transmission frequency you are interested in out of the jumbled cacophony of the radio spectrum.

As we've seen, a radio wave is fundamentally an oscillation, and the magnetic and electric fields composing it alternate with a particular rhythm or frequency just like the swinging pendulum of a clock. So to tune a radio transmitter or receiver, you need to include a circuit that electrically oscillates with a particular rhythm and resists other closely matching frequencies. You need to harness the power of resonance.

Think of it this way. A child on a swing will oscillate back and forth with a particular frequency, just like any pendulum. If you deliver a series of tiny pushes at the right moments the child will swing higher and higher. But pushing with a different rhythm from this resonant frequency will get you nowhere.

Building a basic oscillator circuit that beats with a fixed rhythm uses a gratifyingly elegant combination of a capacitor and an inductor. A capacitor is made up of two metal plates facing each other, sandwiching a layer of insulation between. Any voltage across the device herds electrons onto one of the plates until it becomes so negatively charged it resists further filling. A capacitor serves as a reservoir of electric charge and can release this in a sudden torrent, such as in the flashbulb of a camera. An inductor coil is essentially an electromagnet, but the effect of an inductor is far more than just attracting metal objects. While resistance resists the flow of current, inductance resists any change in the flow of current. So the capacitor and the inductor both serve as refillable stores of electrical energy: the capacitor in the form of an electric field between its facing metal plates, the inductor as a magnetic field surrounding the coil. Wire these two components opposite each other and the simple loop circuit miraculously comes to life.

As the electron-laden capacitor plate dumps its stored charge, it pushes a current around the circuit and through the inductor to create a magnetic field, until the capacitor plates have equalized. Now the magnetic field around the inductor begins to collapse, but as it does, the shrinking field lines sweep over the coil to induce a current in the wire (the generator effect), and continue to pump electrons to the other capacitor plate—amazingly, the collapsing magnetic field is able to temporarily sustain the very electric current that created it in the first place. By the time the inductor field has diminished back to nothing, the opposite plate of the capacitor has become fully charged, and now pushes the current back in the opposite direction, flowing through the coil again.

The energy flows back and forth between capacitor and inductor in this way, being repeatedly interconverted between electric and magnetic fields, like a pendulum swinging to and fro thousands of times every second—at radio frequencies.

The beauty of this disarmingly simple oscillating circuit is that it wants to tick at only its own natural frequency, and will resist other frequencies. You can change the resonant frequency of this circuit, and so retune your transmitter or receiver, by changing the properties of one of the two components. The capacitor is the easier to adjust: rotating D-shaped metal plates past each other varies their overlap and so the charge that can be stored. The tuning knob on old radio sets was therefore often connected to a variable capacitor in the oscillating circuit. Modern transmitters and receivers can be tuned so finely that the radio spectrum has been thinly sliced like a ham on the delicatessen counter and shared between myriad applications: commercial radio and TV stations, GPS signals, emergency service communications, air traffic control, cell phones, short-range Wi-Fi and Bluetooth, radio-controlled toys, and so on. Indeed, spark gap transmitters are now illegal, as they are such unrefined sources, leaking emission in fat smears across the radio spectrum, that they essentially spam broad regions of neighboring radio bands.

The other crucial elements for audio broadcasts are of course a microphone, to convert sound waves into voltage variations in the transmitter circuit, and earphones or speakers to transform the received electric signals back into sound. In fact, microphones and earphones are basically the same device. Both contain a diaphragm that is free to vibrate to either create or respond to sound waves, fixed to a coil of wire that then moves over a magnet, and so they harness the same reversible electromagnetic effects as motors and generators.

A more sensitive version can be built using a piezoelectric crystal, which has the curious property of generating an electric voltage when it is flexed. Such a sensitive crystal earphone is needed to hear the vanishingly faint output from a cat's-whisker radio detector. Potassium sodium tartrate (or "Rochelle salt," after the hometown of the seventeenth-century apothecary who first created it) works nicely in this respect. This salt can be prepared by mixing hot solutions of sodium carbonate and potassium bitartrate (widely known as cream of tartar), which can be gathered as the crystals that form inside wine fermentation casks.

We can be confident that a rebooting civilization could quickly reattain radio communications from absolute basics, even without deriving the complex electromagnetic equations or having the capability to manufacture precision electronic components. It's already been done in our own recent history.

During the Second World War, both soldiers holed up at the front lines and those imprisoned in POW camps built their own makeshift radio receivers for music or news of the war effort. These ingenious constructions reveal the sheer variety of scavenged materials that can be jury-rigged to create a working radio. Aerial wires were slung over trees, or disguised as clotheslines, and sometimes even barbed wire fences were appropriated for the task. A good grounding was achieved by connecting to cold-water pipes in the POW barracks. Inductors were constructed by winding coils around cardboard toilet rolls, the scavenged bare wire insulated by candle wax, or in Japanese POW camps by applying a paste of palm oil and flour. Capacitors for the tuning circuit were improvised out of layers of tinfoil or cigarette-pack lining, alternating with newspaper sheets for insulation; the wide, flat device was then curled like a jelly roll to make a more compact component.

The earphone is a trickier component to improvise, and so was often salvaged from wrecked vehicles. Rudimentary alternatives were constructed by coiling wire around a core of iron nails, sticking a magnet on the end, and lightly positioning a tin can lid over the coil to vibrate weakly with the received signal.

Perhaps the most ingenious improvisation of all, however, was in creating the all-important rectifier, needed to demodulate the audio signal from the carrier wave. Mineral crystals like iron pyrite or galena were unobtainable on the battlefield, but rusty razor blades and corroded copper pennies were discovered to serve just as well. The blade or coin was fixed to a scrap piece of wood alongside a safety pin bent upright. A sharpened pencil graphite was firmly attached to the point of the safety pin (often by winding spare wire tightly around the two), and the springiness of the arm functioned adequately as a cat's whisker, allowing fine readjustment of the pencil graphite across the metal oxide surface until a working rectifying junction was found.

Crystal radios (as well as rust-and-pencil detectors) are beautiful in their simplicity and don't need to be plugged into an electricity source, as they derive their operating power from the received radio wave itself. But the cat's whisker rectifier is unreliable, and crystal sets can produce only very low-power sound. Surprisingly, the solution to this, and a gateway technology for a whole range of other advanced applications, is related to another feature of modern civilization—the light bulb.

Just like a light bulb, a vacuum tube consists of a hot metal filament within a glass bubble, but the vacuum tube, most important, also includes a metal plate around the filament, and the interior is evacuated

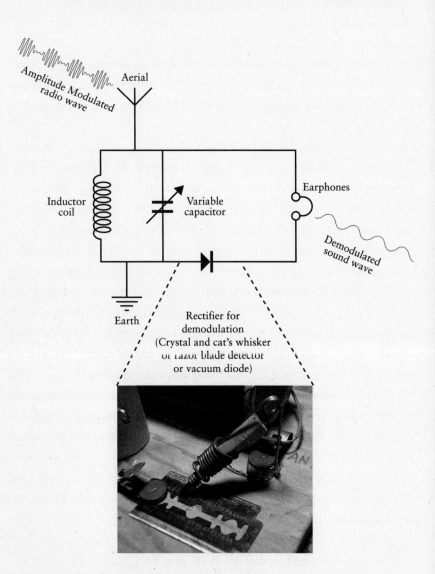

Amplitude Modulated radio wave

Aerial

Inductor coil

Variable capacitor

Earphones

Demodulated sound wave

Earth

Rectifier for demodulation (Crystal and cat's whisker or razor blade detector or vacuum diode)

WIRING DIAGRAM FOR A SIMPLE RADIO RECEIVER (TOP) AND A RAZOR-BLADE RECTIFIER COMMON IN POW RADIOS (BOTTOM).

to a very low pressure. When the filament is heated white-hot, electrons boil off the metal and form a cloud of charge around the wire. This is known as thermionic emission and is what underlies the functioning of X-ray machines, fluorescent lights, and old-style television and computer screens. If the plate is more positively charged than the filament, these freed electrons are attracted over and a current flows through the device. But current can never flow the other way, as the metal plate is not heated to give off electrons, so such a "diode" (with two metal contacts or electrodes) works like a check valve, permitting current in only one direction. Employing very different physics, the thermionic valve therefore exhibits an identical functionality to the crystal detector and can be put to immediate use as the rectifier in radio receivers. But the crucial innovation, enabling a whole new capability, comes from a simple adornment to the diode.

If you take a standard vacuum-tube diode and add a wire spiral or mesh between the hot filament and the metal plate, you can achieve something fantastic. This three-element device is called a triode, and by tweaking the voltage you apply to the mesh you can influence the current flowing through the tube. Applying a slight negative voltage to the control grid begins to repel the electrons boiling off the filament and streaming to the metal plate; increase the negative bias further and the flow is restricted even more—it's like pinching a drinking straw to control how much fluid passes up it. Crucially, the triode allows you to use one voltage to control another. But the ingenius application of the arrangement is that tiny variations in the small control-grid voltage can cause large variations in the output voltage. You have amplified the input signal.

This function is unachievable with crystals and can be used to amplify the weak received signal to power speakers and fill a room with sound. It also enables you to generate a pure-frequency electrical oscillation perfect for a narrow-band carrier wave and conveniently imprint the carrier wave with sound modulation. These are all crucial applications for mainstream radio communications, but just as usefully, vacuum tubes can be used like a switch, far faster than a mechanical lever, to control the course of electricity. Connecting together a large network of these vacuum tubes so that the switches control one another allows you to run mathematical calculations and even construct fully programmable electronic computers.[3]

---

3 Modern electronics have moved beyond power-hungry vacuum tubes, and now exploit the domain of properties of semiconductor materials: vaccum-tube rectifiers are replaced by solid-state diodes, and the voltage-controllable behavior of the triode is reproduced by the silicon transistor. The monolith to miniaturization that is the smartphone in my pocket contains *trillions* of transistors, each of them functionally identical to a warmly glowing vacuum tube.

CHAPTER 11

ADVANCED CHEMISTRY

I wouldn't mind if consumer culture went *poof!*
overnight because then we'd all be in the same
boat and life wouldn't be so bad, mucking about
with chickens and feudalism and the like. But . . .
If, as we were all down on earth wearing rags and
husbanding pigs inside abandoned Baskin-Robbins
franchises, I were to look up in the sky and see a
jet . . . I'd go berserk. I'd go crazy. Either *everyone*
slides back into the Dark Ages or *no one* does.

DOUGLAS COUPLAND, *Shampoo Planet* (1992)

THROUGHOUT THIS BOOK we've looked at a few simple ways that you can convert one substance into another. While these transformations between substances with very different appearances may seem like magic at first, with a little effort you can come to understand the behavior of different chemicals, spot patterns in the way they interact with one another, come to predict what will happen in a reaction, and then, ultimately, wield that power of knowledge to control what happens in a complex set of reactions to deliver exactly the outcome that you want.

Later in this chapter we'll explore how a more advanced civilization, which has secured a stable footing over generations of recovery since the Fall, will be able to employ more complex, industrial processes to provide for its requirements; the rudimentary methods we've covered already for producing soda will get you only so far. But first, let's take a look at how electricity can be used to extract several crucial commodities for the rebooting civilization, and how it helps us to explore the startling order that underlies the chemical world.

ELECTROLYSIS AND THE PERIODIC TABLE

We've seen how mastering the generation and distribution of electricity offers a fantastic power source for a multitude of the functions of a recovering civilization and enables communication across vast distances. But the first actual implementation of electricity in our history, and an application that you will also find invaluable early in a reboot, is using electricity for tearing apart chemical compounds to liberate their constituents: electrolysis.

For example, if you shunt a current through a brine (sodium chloride) solution, you'll be able to collect hydrogen gas bubbling off the negative electrode from the splitting of water molecules, and chlorine gas from the positive. Hydrogen can be used to fill airships and is a raw ingredient for the Haber-Bosch process (to which we'll come later in this chapter), whereas chlorine is valuable for creating bleaches that you will need to make paper and textiles. And if you're a little bit clever with the setup, you'll also be able to extract sodium hydroxide (caustic soda) building up in the electrolyte fluid, which, as we've seen earlier, is a fabulously useful alkali. Electrolysis of pure water (with a little sodium hydroxide added to help it conduct electricity) will yield oxygen and hydrogen.

Aluminum, too, can be teased from its rocky ore by electrolysis— it's too reactive to be smelted using charcoal or coke. It is the most abundant metal in the Earth's crust and a major constituent of one of the earliest materials to be employed by humanity—clay. Yet it was prohibitively expensive until the development of an effective method for melting and electrolyzing its ore in the late 1880s.[1] Luckily, a recovering society will not immediately need to purify the metal afresh. Aluminum is so fantastically resistant to corrosion that it will remain uncorrupted for centuries after the apocalypse and can be recycled by melting at the relatively low temperature of 660°C, using the rudimentary furnace we encountered on page 132.

By employing electrolysis, you'll be able to synthesize several substances useful for civilization, leapfrogging past less effective chemical methods that were employed over the centuries. Moreover, electrolysis will also help with your scientific exploration of the world: it decomposes compounds to retrieve the pure building blocks of all substances—the elements. In 1800, for example, electrolysis conclusively demonstrated that water is not an element at all, but a compound of hydrogen and oxygen. And within eight years another seven elements were isolated by electrolysis: potassium, sodium, calcium, boron, barium, strontium, and magnesium. The first three of these were discovered by using electricity to break down commonplace compounds we have used frequently in this book: potash, caustic soda, and quicklime, respectively. And not only is electrolysis a crucial technique for isolating previously unknown elements: the process also demonstrates that the bonds holding atoms together in compounds are themselves electromagnetic in nature.

If you consider the interactions of the different elements, how they tend to behave in reactions with one another—their personalities—

---

1 In the second half of the nineteenth century, the French emperor Napoleon III hosted a banquet with aluminum cutlery to impress his most distinguished dinner guests, laying out the aluminumware rather than silverware. Bizarrely, it was simultaneously the most common and the most precious metal on the planet. But with the development of a suitable flux and the use of electrolysis for mass production, aluminum slumped from the prestige of royal dinner sets to the indignity of drink cans discarded by the millions.

you'll become aware of one resounding, fundamental truth: the elements aren't solitary but naturally fall into clusters with similar behaviors, like families. The discovery of this pattern gives structure to the chemical universe, in the same way that the realization of morphological similarities and thus relatedness between living organisms brings order to the biological world. Sodium and potassium, for example, are both violently reactive metals that form alkaline compounds, such as the caustic soda and potash you can electrolytically isolate them from; and chlorine, bromine, and iodine all react with metals to form salts. If you now sort the known elements into an array, lining up those with similar behavior in the same column to represent the underlying repeating pattern, you create the periodic table of the elements.

The modern periodic table is a colossal monument to human achievement, as impressive as the Egyptian pyramids or any of the other wonders of the world. Far more than just a comprehensive list of different elements that chemists have identified over the years, it is a way of organizing knowledge that allows you to predict details about what you have not yet found.

For example, when the Russian chemist Dmitri Mendeleev first assembled a periodic table in 1869 of the 60-odd elements then known, he found gaps in the brickwork—placeholders corresponding to missing substances. But the brilliant thing about the arrangement, where the elements are placed according to their properties, is that it enabled him to predict precisely what these hypothetical elements would be like—such as eka-aluminum, the missing piece in the table immediately below aluminum. Even though this hypothetical stuff had never been seen or touched, based purely on its location within the array you can predict that it would be a shiny, ductile metal, with a particular density, and that it would be solid at room temperature but melt at an unusually low temperature for metals. A few years later, a Frenchman discovered a new element in an ore and named it gallium, after the old name for his homeland. But it soon became clear that this was the missing eka-aluminum anticipated by Mendeleev, and that his prediction for the melting point was spot-on: gallium turns from solid to liquid at a temperature of 30°C—the metal literally melts in your hand.[2]

This simple truth about the patterns inherent in the elements will help structure your own post-apocalyptic investigations into the makeup of matter and how to best exploit the different properties offered by natural substances. Let's turn now to build on the lessons from Chapters 5 and 6 and take a look at two useful applications of slightly trickier chemistry—explosives and photography.

EXPLOSIVES

You might think that explosives are exactly the sort of technology you would want to leave out of a manual for rebooting civilization, to prolong peaceful coexistence for as long as possible. It is certainly true that explosives can be turned to warmongering (or defensive) ends, and historically their chemistry has developed in tandem with the metallurgy required to safely contain and direct the blast for reliable cannon or firearms. But the peaceable applications are arguably far more crucial for a recovering civilization: explosives are enormously helpful in rifles for hunting, as well as to break open rock faces for quarrying and min-

---

2 Indeed, since the 1930s we've been going one step further, and filling additional rows at the bottom of the periodic table with elements that do not exist naturally, but were technologically created—atoms with a nucleus so swollen with protons and neutrons that they are acutely unstable and disintegrate again almost immediately in a flurry of radioactivity. Through our own history, therefore, we have not only cooked up new materials—ceramics like glass or metallic mixtures such as steel alloys—or novel molecules like the organic polymers of plastics, but learned how to transmute elements themselves, achieving the dream of the alchemists. And with dedication, a civilization following in our footsteps would be able to accomplish the same.

ing, and for blasting tunnels and canals. And perhaps most important in a post-apocalyptic world will be the demolishing of dilapidated and unsafe high-rise buildings to cannibalize their structural components and clear the land for redevelopment as the emergent civilization expands back into long-deserted quarters. In any case, scientific knowledge itself is neutral: it is the purpose to which it is applied that is either good or evil.

To create an explosion—a rapidly expanding pulse that assaults your eardrums, shatters a rock face, or pushes over a building—you need to suddenly create a bubble of very-high-pressure air in a small space. And the best way to accomplish that is with a frenzied flurry of chemical reactivity that converts solid substances into hot gases, which take up far more room and so quickly expand outward from the reaction point. A modern rifle, for example, contains roughly a sugar cube's worth of powder in the charge behind the bullet, but when this is triggered it reacts with itself blindingly quickly to create a ball of gas about the size of a party balloon. The rapid expansion within the claustrophobic confines of a narrow rifle barrel is what creates the enormous force that hurls the bullet at about the speed of sound.

You can get solid fuels to explode by grinding them into a fine powder so that the air has access to a much greater area to accelerate combustion; coal dust and flour burn extremely vigorously (and explosions can occur even at grain elevators). An even better solution is to remove the necessity for getting oxygen from the air, and instead provide plenty of oxygen atoms already in close proximity to the fuel for rapid combustion. A chemical that generously supplies oxygen atoms—or, more generally speaking, is hungry to accept electrons off other chemicals—is called an oxidizing agent or oxidant.

Ironically enough, the earliest explosive to be developed in history was first formulated by ninth-century Chinese alchemists seeking an elixir for immortality: black powder. Gunpowder consists of charcoal—the fuel or reductant—and saltpeter (now termed potassium nitrate), the oxidant, ground and mixed together. Sprinkling in some yellow elemental sulfur as a third ingredient changes the end products of the reaction and results in far more energy being left over for the concussive whump. An optimized gunpowder recipe is to mix equal parts of saltpeter and sulfur to six parts of charcoal fuel: a chemical cocktail taut with latent energy poised to burst out.

The nitrate ingredient of gunpowder calls for a bit of nifty chemical wheeling and dealing. Historically, the source of nitrates for explosives as well as fertilizers was very humble: a well-matured pile of manure contains hordes of bacteria that have acted to convert nitrogen-containing molecules into nitrates, and you can get these out by exploiting the fact that similar compounds have differing abilities to dissolve in water. It is a fact of chemistry that all nitrate salts are readily water-soluble, and that hydroxide salts are often insoluble. So, soak a few buckets of limewater (calcium hydroxide; see Chapter 5) through a dung pile, and most of the minerals will stay trapped inside as insoluble hydroxides, while the calcium will pick up the nitrate ions and drain out. Collect this fluid and stir in some potash. The potassium and calcium will swap partners to create calcium carbonate and potassium nitrate. Calcium carbonate doesn't dissolve in water—it's the compound making up limestone and chalk, and the white cliffs of Dover certainly aren't vanishing with every wave—but potassium nitrate can. So filter out the chalky white precipitate before boiling away the water to yield crystals of saltpeter. A good test that your isolation has been successful is to soak some of the solution onto a strip of paper and let it dry—if you've got potassium nitrate it will burn with a fizzing, sparkling flame.

The chemistry for extracting saltpeter is straightforward enough; the trouble is in finding enough sources of nitrates to use as feedstocks for the process as the demands of your recovering civilization grow.

Suitable mineral deposits are found only in very arid environments (saltpeter is readily soluble and thus easily washed away) such as the Atacama Desert in South America, and bird guano is also very rich. The use of nitrates in both fertilizers and explosives meant that they had become a crucial commodity by the end of the nineteenth century, and wars were fought over tiny barren islands for the bird shit they were encrusted in. We'll take a look later in this chapter at how to release your developing civilization from the constraints imposed by nitrogen starvation.

While gunpowder supports rapid combustion by intermingling fuel and oxidant powders snugly together, there is an even better way to ensure a more vigorous reaction and thus a more powerful explosion: combining the fuel and the oxidant into the same molecule. Reacting many organic molecules with a mixture of nitric and sulfuric acids (see Chapter 5) serves to oxidize them, tacking on nitrate groups to the fuel molecule. For example, oxidizing paper or cotton (which are both sheets of plant cellulose fibers) with nitric acid produces the heartily flammable nitrocellulose—flash paper or guncotton.

Another explosive more potent than gunpowder is nitroglycerin. This clear, oily explosive is made by the nitration of glycerol, an off-shoot of the production of soap, as we saw in Chapter 5, but it is disastrously unstable and liable to blow up in your face at the slightest provocation. The solution that Alfred Nobel found to stabilize its destructive potential was to soak the shock-sensitive nitroglycerin into wads of absorbent material like sawdust or siliceous clay—creating sticks of dynamite. (It was the fortune from this invention that Nobel used to found the famous prizes for contributions to humanity in the sciences, literature, and peace.)

The production of powerful explosives, therefore, relies on nitric acid as a potent oxidizing agent, and this same acid is also required for photography and the capturing of light using silver chemistry.

Photography is a wonderful technique—a way of harnessing light to record an image, capturing an instant in time and preserving it for eternity. A holiday picture can trigger vivid reminiscences even decades later, but photography records the visual world with far greater fidelity than memory can ever offer. Yet beyond drunken party snapshots, family portraits, or breathtaking landscapes, the incomparable value of photography over the past two hundred years has been in presenting what the eye cannot see. It represents a key enabling technology across numerous fields of science, and will be vital in accelerating the reboot. Photography allows investigators to record events and processes that are exceedingly faint or occur over timescales too rapid or too slow for us to perceive, or at wavelengths invisible to us. For example, photography offers extended exposure times to soak up feeble light over far longer periods than the human eye can offer, allowing astronomers to study a multitude of dim stars and resolve faint smudges into detailed galaxies and nebulae.[3] Photographic emulsions are also sensitive to X-rays and so allow you to create medical images for examination of the body's interior.

---

3 You can also use a camera to demonstrate the existence of a prior, technologically advanced civilization on Earth even far into the future. A photo taken of the night sky toward the celestial equator (90 degrees from the pole, see Chapter 12) with an exposure of a minute or two will smudge all the stars into curved streaks with the Earth's spin. But occasionally you'll also spot something very curious: pinpoints of light that haven't been smeared at all. These objects apparently fixed in the sky must just so happen to be moving at exactly the same rate that the planet turns—artificial things that were deliberately placed around the Earth in this particular configuration. These are the geostationary satellites ringing the equator at a special distance where the orbital period is exactly one day; such satellites remain fixed above the same point on the Earth's surface and so make good communications relays. Their orbit is also exceedingly stable, and long after all of our cities and other artifacts have crumbled to dust and been buried, they will remain monuments to our technological civilization in the pristine environment of space, easy to spot if you know how.

The crucial chemistry behind photography is simple enough: certain compounds of silver darken in sunlight and so can be employed to record a black-and-white image. The trick is to create a soluble form of silver that can be spread evenly in a thin film, but then convert it into an insoluble salt that sticks on the outside surface of your photographic medium and doesn't get washed away again.

First, coat a sheet of paper with egg whites containing some dissolved salt, and allow it to dry. Now dissolve some silver in nitric acid, which will oxidize the metal to soluble silver nitrate,[4] and spread the solution over your prepared paper. The sodium chloride will react to create silver chloride, which is both light-sensitive and insoluble, and the egg albumin will prevent the photographic emulsion from soaking into the paper fibers. If you scavenge in the post-apocalyptic world, a single solid-silver teaspoon will contain enough of the pure element to produce over 1,500 photographic prints.

When light rays hit this sensitized paper, they provide the energy to liberate electrons in the grains and so reduce the silver chloride back to metallic silver. Large lumps of silver, such as a polished platter, have a bright luster, but a speckle of tiny metallic crystals scatters the light instead and so looks dark. On the other hand, areas of the sensitized sheet not exposed to light remain the white of the paper behind. The key follow-up step after the exposure is to kill this photochemical reaction and so stabilize the captured shadows. Sodium thiosulfate is the fixing agent still used today and is relatively easy to prepare. Bubble sulfur dioxide gas (page 121) through a solution of soda or caustic soda,

---

4 While we're on silver chemistry, it's worth mentioning another crucial capability: that of creating a mirror, indispensable beyond mere vanity as a critical component of high-powered telescopes or the sextant for navigation. Mix alkaline ammonia solution (see Chapter 5) with silver nitrate and a little sugar, and then pour it over the back of a clean piece of glass. The sugar reduces the silver back to pure metal and so deposits a thin shiny layer directly onto the glass surface.

then boil with powdered sulfur and dry for crystals of "hypo" (a nickname derived from its old name, hyposulfite of soda).

Using a lens set into the front of a light-tight box to project an image onto sensitive paper on the back wall produces a photographic camera, but even in bright sunshine it can take many hours for this rudimentary silver chemistry to take a "photo." Luckily, you can increase the sensitivity of your camera enormously with a developer—a chemical treatment that completes the transformation of partially exposed grains, reducing them entirely to metallic silver. Ferrous sulfate works well, and can be synthesized easily enough by dissolving iron in sulfuric acid. And as the chemical proficiency of the post-apocalyptic society improves, in place of chlorine salt you can substitute that of one of its atomic siblings, iodine or bromine, which produce far more light-sensitive photographic emulsions.

However, the fact that light exposure turns the photosensitive grains dark with silver metal, whereas shadows in the scene remain pale, means that the photo comes out tonally reversed from what your eye sees—you'll get a "negative." There is no fast-acting chemical reaction that produces a permanent positive image—no initially black substance that rapidly bleaches in sunlight—and so photography is burdened with this negative outcome. The necessary conceptual leap is to realize that if this reversed, negative image is created in the camera on a transparent medium, all that is needed is a second stage of printing out using the negative as a mask on top of sensitized paper, so that the pattern of highlights and shadows reverses again back to normal. The wet collodion process uses guncotton dissolved in a mixture of ether and ethanol solvents—all substances we have already come across in this book—to produce a syrupy, transparent fluid. It's perfect for coating a glass plate with photochemicals, and then exposing and developing the image before it dries into a tough, waterproof film. And if instead you use gelatin (boiled out of animal bones, as we saw in Chapter 5), it is possible to create a dry plate that is even more photosensitive and allows much longer exposure times.

Photography is a fantastic example of a novel application created by the fusion of several preexisting technologies, and using relatively straightforward materials and substances. Build a fireclay-lined kiln to produce your own glass by melting silica sand or quartz with a soda ash flux. Take one dollop to grind into a focusing lens and another to flatten into a rectangular pane for a negative plate, and draw on your papermaking skills to produce smooth prints. The chemistry underlying photography uses the same acids and solvents marshaled time and time again in this book, and you can take a primitive photo using substances derived from a silver spoon, a dung heap, and common salt. Indeed, if you fell through a time warp back to the 1500s you'd readily be able to source all the chemicals and optical components you needed to build a rudimentary camera, so you could show Holbein how to take a photograph of King Henry VIII rather than create an oil painting.

Filling in the periodic table of the elements, exploiting explosives, and using photography as a tool for rediscovery will all be important activities for a civilization restarting after the Fall. But as the post-apocalyptic society recovers and begins to flourish, it will need ever-increasing amounts of the basic substances we've discussed throughout this book. And to meet these demands, civilization will need to develop the more advanced processes of industrial chemistry.

INDUSTRIAL CHEMISTRY

We often hear about the Industrial Revolution and the innovation of ingenious mechanical contraptions for alleviating the toil of humankind, thereby greatly accelerating the pace of progress and transforming eighteenth-century society. But the transition to an advanced civilization is as much about the invention of chemical processes for the large-scale synthesis of the necessary acids, alkalis, solvents, and other substances critical to the running of society as it is about machinery for automating spinning and weaving and building roaring steam engines.

Many of the vital necessities of civilization that we've covered in this book rely on the same reagents to drive the transformations from raw materials gathered from the environment to the required commodities or products. And as the post-apocalyptic population swells over the generations of recovery, your ability to meet the demand for these critical substances by using the rudimentary methods we've looked at so far will become limiting, threatening to stall further progression.

We'll focus here on the creation of two substances that became grievous choke points in our own history: soda in the late 1700s and nitrate in the late 1800s. Securing an adequate supply of both of these will also inevitably become critical for a rebooting post-apocalyptic society. So how can a recovering civilization release itself from the restrictions of relying on ashes for soda or manure for nitrates? Let's start with the large-scale synthesis of soda, which formed the very beginnings of industrial chemistry in our history.

As we've seen, soda ash—sodium carbonate—is a vitally important compound used in a vast number of different activities throughout society. It's indispensable as a flux for melting sand to produce glass (more than half of the global production of sodium carbonate today is used for glassmaking), and when converted to caustic soda (sodium hydroxide) it's the best agent for driving the central chemical reactions in creating soap and separating the plant fibers needed for making paper. Glass, soap, and paper are central pillars of civilization, and since the Middle Ages we have relied upon a constant, cheap supply of alkali.

Traditionally, alkali was provided by potash from burned timber, and by the eighteenth century, widespread deforestation of Europe meant that potash came to be imported from North America, Russia, and Scandinavia. For many applications, however, soda ash is preferred (the caustic soda made from it is a far more potent hydrolyzing agent than caustic potash), which was produced in Spain by burning the native saltwort plant, and along the Scottish and Irish coastlines from kelp washed ashore by storms. Sodium carbonate was also mined from dried lake-bed deposits of the mineral natron in Egypt. But by the second half of the eighteenth century, as the population and economy grew, the demand for soda began to outstrip the supply from these natural sources, as will inevitably happen again with a recovering post-apocalyptic society. Common sea salt and soda ash are chemical cousins,[5] so can you convert an essentially limitless substance into an economically crucial commodity?

One simple two-step operation, developed by the eighteenth-century French chemist Nicolas Leblanc, is to react the salt with sulfuric acid and then roast the product with crushed limestone and charcoal or coal in a furnace at around 1,000°C to form a black ashy substance. The sodium carbonate you are interested in is soluble in water, so you can soak it out using exactly the same technique you would use when working with seaweed ashes. However, while this Leblanc process is a readily achievable way of morphing salt into soda, releasing you from the limitations of burned plants or mineral deposits, it is grossly inefficient and churns out noxious waste products.[6] So ide-

---

5 In modern nomenclature, we would say that common sea salt (sodium chloride) and soda ash (sodium carbonate) are both chemical salts of the same base (sodium hydroxide, known traditionally as caustic soda).

6 In the early nineteenth century these were simply dumped as toxic waste: insoluble heaps of dark ashy calcium sulfide were piled up in the fields surrounding the soda works, and clouds of hydrogen chloride poured out of tall chimneys, causing great damage to surrounding plant life. In 1863 Britain passed the Alkali Act, which prohibited the venting of hydrogen chloride—the first modern legislation against air pollution. The immediate response of the soda plants was to scrub this soluble gas by sprinkling water down the inside of their chimneys and discharge the resultant hydrochloric acid straight into the nearest river, deftly sidestepping the legislation by turning air pollution into water pollution!

ally the rebooting civilization would want to be able to leapfrog straight over the easy but wasteful LeBlanc process to a more efficient system.

The Solvay process is slightly more intricate, but ingeniously employs ammonia to close the loop: the reagents it uses are recycled back within the system, minimizing wasteful byproducts and thus also pollution. The chemical reaction sitting at the core of the Solvay process is this: when a compound known as ammonium bicarbonate is added to strong brine, the bicarbonate ion swaps over onto the sodium to form sodium bicarbonate (identical to the rising agent used in baking), which can then simply be heated to change to soda ash. The first step in achieving this is to pass the strong brine through two towers, with first ammonia gas and then carbon dioxide bubbling up through them to dissolve in the salty water and combine to make the crucial ammonium bicarbonate. The swapping reaction occurs with the salt, creating sodium bicarbonate, which doesn't dissolve and so settles as a sediment to be collected. The ammonia is the key ingredient for this stage, as it keeps the brine nicely alkaline and so ensures that the bicarbonate of soda can't dissolve, neatly separating these two salts.

The carbon dioxide needed for this initial step is baked out of limestone in a furnace (in exactly the same way we saw in Chapter 6 for burning lime for mortar and concrete production). The quicklime left behind is itself added to the brine solution after the soda has been extracted and regenerates the ammonia bubbled in originally, ready to be used again. So overall, the Solvay processes consumes only sodium chloride salt and limestone, and alongside the valuable soda produces just calcium chloride as a byproduct, which itself is used for spreading as a de-icing salt on wintry roads. This elegantly self-contained system, cleverly recycling the key ammonia as it goes and built using only fairly rudimentary chemical steps, is still the major source of soda worldwide today (except in the United States, where a large deposit of the sodium carbonate mineral trona was discovered in Wyoming in the 1930s). And for a recovering civilization the Solvay process presents a marvelous

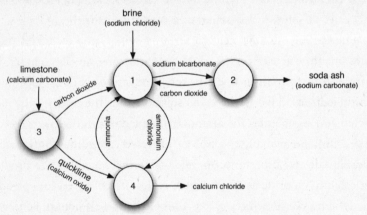

brine
(sodium chloride)

limestone
(calcium carbonate)

sodium bicarbonate

soda ash
(sodium carbonate)

carbon dioxide

carbon dioxide

1

2

3

ammonia

ammonium chloride

quicklime
(calcium oxide)

4

calcium chloride

A SODA PLANT IN NEW YORK IN THE LATE NINETEENTH CENTURY, OWNED BY THE SOLVAY PROCESS COMPANY (TOP). THE FOUR STEPS OF THE SOLVAY PROCESS FOR ARTIFICIALLY SYNTHESIZING SODA (BOTTOM). THE RECYCLING OF AMMONIA SITS AT THE HEART OF THIS CRITICAL CHEMICAL PROCESS.

opportunity to leapfrog over less efficient and noxiously polluting alternatives for producing vital soda to form the foundation of a post-apocalyptic chemical industry.

The Solvay process converts an abundant source of the element sodium, common salt, into the crucial alkaline compound soda. But before too long an advancing civilization will run into problems of supply limitation of another critical commodity. One of the most fundamental chemical processes for all of us alive today involves the element nitrogen—and another miraculous transformation of a common base substance into something vitally valuable.

In terms of the sheer number of people it directly affects every day, the most profound technological advance of the twentieth century was not the invention of flight, antibiotics, electronic computers, or nuclear power, but the means to synthesize a humble, foul-smelling chemical: ammonia. As we've seen throughout this book, ammonia and the related (and thus chemically interconvertible) nitrogenous compounds nitric acid and nitrates are foundation stones of the chemistry underpinning civilization. Nitrates are absolutely vital for the production of both fertilizers and explosives, but by the closing years of the nineteenth century, the industrialized world was running out: demand was starting to exceed supply, and America and European nations became concerned not just about ensuring munitions for their armies, but about fundamentally providing enough food to keep their citizens alive.

For millennia, the response to an expanding population had been simply to clear more ground for cultivation. Once you've hit the limit of available land, though, the only solution to feed the multiplying mouths is to increase the yield of crops from the same cultivable area. As we saw in Chapter 3, returning manure to the soil and planting legumes are both effective. But when a population reaches a certain limit—a capacity crowd, if you like—civilization inescapably hits a snag. You cannot produce more manure from livestock, as animals would need to be fed plants grown on the land in the first place; and you can't sow more fields with legumes, as that decreases the land available for growing cereal crops. You've hit the carrying capacity of organic agriculture.

The only recourse is to inject external nitrogen from outside of the agricultural loop. Through the nineteenth century, Western agriculture relied heavily upon imported guano and on saltpeter mined from the Chilean desert. But these sources rapidly became depleted, and the president of the British Association for the Advancement of Science, Sir William Crookes, warned in 1898 that "we are drawing on the Earth's capital, and our drafts will not perpetually be honored" (a caution we would be wise to heed today as our civilization's voracious appetite for crude oil and other natural resources threatens to exhaust them). The world we leave behind will already have been stripped of these natural deposits of nitrate, and a maturing post-apocalyptic civilization will slam into this wall soon enough.

The planet's atmosphere is rich with nitrogen gas—it makes up almost four-fifths of every breath you take—but it's recalcitrantly unreactive. The two atoms of nitrogen are locked firmly together by a triple bond; in fact, nitrogen gas is the least reactive diatomic substance known. This makes it very difficult to convert nitrogen gas into accessible forms—to "fix" it. By the end of the nineteenth century it had become clear that figuring out how to fix nitrogen was critical to the progress of civilization itself—that chemistry had to come to the rescue of humanity.

The solution, discovered in 1909 and still used today, is known as the Haber-Bosch process. On the face of it, the process appears beguilingly simple. Nitrogen, the most common gas in the Earth's atmosphere, and hydrogen, the most abundant element in the entire universe, are the only raw materials and are mixed together in a one-to-three-parts recipe within a reactor and combined to form NH_3—ammonia. Nitrogen can be simply sucked out of the air, and today hydrogen is made from methane natural gas, but it can also be gathered from the electrolysis of water. Coaxing nitrogen into participating requires severing the sturdy bonds clamping the twin atoms together, and that requires a catalyst. A porous form of iron, with potassium hydroxide (the caustic potash we covered on page 115) as a promoter to increase its effectiveness, works well for hustling this reaction along. The reaction is never complete, so the gases are cooled for the desired product to condense as ammonia rain to be drained off and stored, and the as-yet-unreacted gases are recycled back through the reactor repeatedly until virtually everything is successfully transformed. But, as with many things, the devil is in the details, and the Haber-Bosch process is actually pretty tricky to pull off.

Many chemical reactions are essentially unidirectional: they are a one-way street of reactants recombining into products. For example, in a burning candle the waxy hydrocarbon molecules are oxidized by the combustion process into water and carbon dioxide, but the reverse transformation would never occur spontaneously. Other chemical processes, however, are reversible reactions, and the two opposing conversions run in both directions simultaneously. "Reactants" are transformed into "products," but these are being converted back again at the same time. The conversion between a nitrogen-hydrogen mixture and ammonia is one such reversible process, and to tip the balance toward the desired compound, you need to carefully arrange the conditions within the reactor. For producing ammonia, this means running at a high temperature (around 450°C) and crushingly high pressure (around 200 atmospheres). And these extreme conditions for the reactor and pipework are why the Haber-Bosch process is so troublesome to run. Far more so than the other crucial processes we've looked at that require the heat of a furnace—such as glassmaking or metal smelting—pulling off nitrogen fixation is a feat of accomplished engineering. If your post-apocalyptic society can't salvage a suitable reactor vessel, you're going to need to learn how to construct your own industrial pressure cooker.

Persuading nitrogen gas to combine with hydrogen for ammonia is just the first step, though. Once the nitrogen has been fixed, you'll want to convert it to a more generally useful chemical: nitric acid. The ammonia is oxidized in a high-temperature converter—not just a furnace, but a vessel that essentially burns the ammonia gas itself as a fuel, using a platinum-rhodium catalyst. This is actually the same alloy as in the catalytic converter stuck in the exhaust pipe of cars to reduce pollution emissions, and so should be relatively easy to scavenge in the aftermath. The nitrogen dioxide produced is then absorbed into water to create nitric acid.

Neither of these products—ammonia or nitric acid—can be poured directly onto a farmer's field to help boost crop growth: the first is too alkaline, the second too acidic. But if you simply mix the two together, they neutralize and form the salt ammonium nitrate, and this makes a fantastic fertilizer, as it packs a double dose of accessible nitrogen. As we saw in Chapter 7, ammonium nitrate is also useful in medicine, as it decomposes to release the anesthetic nitrous oxide. It is a powerful oxidizing agent, too, and so can be used to make explosives.[7] So for a post-apocalyptic society maturing into an industrialized civilization, the Haber-Bosch process will liberate you from dependence on collecting animal manure or bird guano, soaking timber ashes, or digging saltpeter mineral deposits for your vital nitrate supply, and instead enable you to mine the virtually limitless reservoir of nitrogen in the atmosphere.

Today, the Haber-Bosch process pumps out around a hundred million tons of synthetic ammonia every year, and fertilizer made from it sustains one-third of the world's population—around 2.3 billion hun-

---

7 More than two tons of ammonium nitrate fertilizer were packed by Timothy McVeigh into the back of a truck for the Oklahoma City bombing, and one of the world's largest non-nuclear explosions occurred in 1947 when a fire caused a ship carrying more than 2,000 tons of the compound to detonate in the port of Texas City.

gry mouths are fed by this chemical reaction. And since the raw materials in the food we eat become assimilated into our cells, about half of the protein in our bodies is made from nitrogen fixed artificially by the technological capability of our own species. In a way, we're partially industrially manufactured.

CHAPTER 12

TIME AND PLACE

Men go and come, but earth abides.

ECCLESIASTES 1:4

The ideas ruins evoke in me are grand. Everything comes to nothing, everything perishes, everything passes, only the world remains, only time endures.

DENIS DIDEROT, *Salon of 1767*

IN THE LAST CHAPTER we built up to some pretty complex industrial chemistry, suitable for supporting the requirements of a burgeoning post-apocalyptic society generations into its recovery after the Fall. Now I want to go right back to basics. What could survivors do to work out from absolute scratch the answers to two crucial questions: "What time is it?" and "Where am I?" This is far from a frivolous exercise: being able to trace your passage through time and through space are critical capabilities. The first allows you to measure the passing of time during the day and to track the days and seasons, a prerequisite for successful agriculture. We'll see what observations you can make to reconstruct the calendar surprisingly accurately, and, if you wanted to, even work out far in the unknown future what year it is, the classic question that falls from the hero's lips in every time-travel film. The second is important for allowing you to trace your location around the globe in the absence of recognizable landmarks. This is crucial for working out where you are in relation to where you want to be, and enables you to navigate for trade or exploration.

Let's look at time first.

TELLING TIME

One of the most fundamental functions of any civilization is to be able to track your passage through the seasons, to know the best moments to sow and then harvest and so prepare for the inescapable approach of the deadly winter or dry season. And as a society becomes more sophisticated and its routines more rigorously structured, telling time within the day becomes increasingly important. Clocks are indispensable for regulating the duration of different activities and synchronizing civic life. Everything from the working hours of tradesmen to the start and end of market, and in religious societies the moment to congregate at the place of worship, is choreographed to the beat of the hour.

In principle, you can meter time by capitalizing on any process that proceeds at a constant rate. A plethora of different methods have been used historically, and will be useful in the early stages of a reboot if no clocks survived. These include the regular dripping of a water clock, with time indicated as a graduated line down the side of either the reservoir or the receptacle; or the trickle of sand or other granular material through a small hole; or the remaining level in an oil lamp; or a scale marked down the side of a tall candle.

The water clock and hourglass work on similar principles of gravity, but unlike the pressure forcing liquid out the bottom of the water clock, the rate of flow of the hourglass is largely independent of the height of the remaining sand column, and this superior timer became common from the fourteenth century on. But while an hourglass can measure duration, it can't itself tell you the time of day (without a rigorous system of repeatedly inverting timers from the moment of dawn). So how can you determine from first principles what the time is?

The structure of our hectic modern lives is dictated today by our wall clocks and daily planners, but these are both no more than formalizations of the primordial rhythms of the planet we live on. On the timescale of our everyday experience, the natural rhythms of the Earth play out too slowly for most of us to be aware of more than the regular beat of day and night or the more gradual cycling of the seasons. If we were able to twist a dial and accelerate the passage of time around us, these planetary periodicities would become much more apparent. (The descriptions below are for a point of view in the northern hemisphere, but if you're in the south the principles are the same.)

With the Sun sliding more rapidly through the sky, shadows swing across the ground, pivoting around the base of the objects casting them. As the Sun races to the west and slips out of view after a cruelly abbreviated sunset, the sky drains color to indigo and the pitch darkness of nighttime descends. The vast spattering of stars across the heavens are not the static points you are used to, but thin lines of light wheeling around the dome of the firmament. They trace out concentric rings nestled within one another, and right in the center, the north celestial pole, no movement is discernible. In the very bull's-eye of this pattern there happens to lie a star, Polaris, or the North Star, around which all others appear to whirl, before the firmament lightens again with dawn.

Next, you notice that the fiery streak of the Sun's trajectory across the sky isn't steady over the weeks, but that the arc rocks gently back and forth. During summer the Sun arcs the highest, providing long, warm days, but during winter it's almost as if the Sun is taking a shortcut, at more northerly latitudes barely hauling itself above the horizon before soon sinking out of view again. The highest and lowest reach of this rocking, where the solar arc seems to slow and stop before swinging back in the opposite direction, is called the solstice (from the Latin for the Sun standing still). The winter solstice (coincident with the summer solstice in the southern hemisphere) is the shortest day of the year, and corresponds to the Sun rising from its southernmost point on the horizon. Ancient astronomical sites like Stonehenge have monuments aligned to the location of sunrise on these special days.[1]

So how can you use these natural rhythms and cycles to determine the time?

To a first approximation, the Sun's journey across the sky as the world spins,[2] and by extension the shifting location of shadows, indicates the time of day. Anyone who has ever tried to stay in the shade of a tree or a beach umbrella will be acutely aware of how it shifts. So if you plant a stick upright in the ground, the rotation of its shadow will indicate the passing of time. This is, of course, the essence of a sundial. The time when the shadow is shortest is midday, or noon. For the most accurate results, the stick (gnomon) should be angled directly toward celestial north, indicated by the polestar, as we saw earlier.

If you can fashion a hemispherical shell to sit around the base of the stick, or a circular arc, the hour lines are simply marked out at regular intervals, as it allows for a direct projection of the celestial sphere onto the curved surface of the sundial. A flat circular sundial is much easier to actually construct, but the marking of the hour lines is more in-

---

1 The parallel streets of Manhattan's grid-like layout are aligned along a bearing of 30 degrees east of celestial north, and twice a year (in late May and mid-July, evenly flanking the summer solstice), this grid behaves like a city-size Stonehenge, with the sun setting right down the centerline of the canyon-like streets.

2 If you need convincing, you can demonstrate that the tracking of the Sun across the sky and the wheeling of the nighttime canopy of stars is caused by our motion, not theirs. Dangle a heavy bob from a long piece of string, indoors and away from any gusts of wind, and carefully set it swinging, straight back and forth without any sideways sway. The swing of "Foucault's pendulum" will seem to rotate around the ground over the course of a day. But the pendulum is suspended in open air, and so there can be no force acting to cause it to twist; in fact, the pendulum is swinging in the same direction all along as the Earth itself turns underneath.

volved because the shadow will move more slowly around noon than in the morning or evening. You can divide the day into as many hours as you like. Our convention of breaking down a full day into two halves of twelve hours each originates with the Babylonians (possibly linked to the twelve signs of the zodiac: the band of constellations that the Sun and planets appear to move through in their trajectory across the sky).

The major revolution in timekeeping in our history, though, and a technology you'll want to aim toward during recovery, is that of a mechanical "clockwork" clock.[3] This is a marvelous contraption that ticks with a rhythmical beat like a heart. Four key components are needed for this action: a power source, an oscillator, a controller, and the clockwork gearing.

The primary part of any mechanism is the power source, and the simplest means for providing this is a weight suspended on a string wrapped around a shaft, so that it turns as the weight descends under gravity. But the crucial problem is how to regulate the release of that stored energy to drive the slow movements of clockwork rather than letting the weight simply drop straight to the ground.

The beating heart of the mechanical clock, the bit that provides the regular timing signals, is called the oscillator. The ideal low-tech solution is a simple pendulum: a swinging mass on a rigid rod. The physical principle you're exploiting is that the period of a pendulum—the time it takes to swing down through a small angle and return to its initial position—is determined by its length. A pendulum will swing with exactly the same beat even as friction and air drag gradually reduce the amplitude of its swing, and it is this incredible regularity that makes it such a useful component for a clock. The third element, the controller,

---

3 These first appeared in late-thirteenth-century monasteries, their chimes calling the monks to prayer. In fact, the crucial mechanisms predate the adoption of clock faces and hands by more than a century (and the minute hand didn't appear for another three hundred years): the earliest clocks couldn't display the time but were ingenious automated systems for ringing bells (indeed, the name derives from the Celtic for bell).

THE KEY COMPONENTS OF A MECHANICAL CLOCK. THE DROP OF THE WEIGHT (BOTTOM LEFT) DRIVES THE GEAR CHAIN, WITH THE ESCAPEMENT (TOP) RELEASING THE TRAPPED WHEEL TO TURN ONE TOOTH AT A TIME AS IT ROCKS BACK AND FORTH. THE ESCAPEMENT IS COUPLED TO THE REGULAR BEAT OF A PENDULUM (NOT SHOWN).

performs the vital job of integrating the timing signal from the oscillator to regulate the power source. The pendulum escapement is a jagged-toothed gearwheel that repeatedly locks and disengages with a two-armed lever that rocks with the pendulum swing. At the top of each swing, the released escapement jolts around one step from the pull of the drive weight, and its angled teeth give a little nudge to keep the pendulum going. So this ingenious arrangement captures the regular impulse of the swinging bob to trickle out the stored energy one tick at a time. The twin demands for a nice long pendulum and high drop for the drive weight dictate the design of many timepieces to look like tall grandfather clocks.

After this, it's a relatively simple affair to design a system of gears that essentially performs a mathematical calculation to adjust the stepwise turn of the escapement to a driven wheel that turns a full circle every 12 hours for the hour hand of a clock face, and a minute hand geared to this at a ratio of 60:1. It is another legacy of the ancient Babylonians that we chop hours into 60 minutes (the name deriving from the Latin *partes minutiae primae*, meaning first small part) and those further into 60 seconds (Latin *partes minutiae secundae*). Pendulum clocks also enable the precise measuring of natural processes and experiments, a development that in our history contributed enormously to the tool kit of investigators through the scientific revolution.[4]

The length of hours indicated by the roving shadow of a sundial varies over the year: a winter hour is shorter than a summer hour. Only on two days of the year are the solar hours all equal: the equinoxes (meaning literally "equal night" because day and night are both 12 hours long).[5] These special days occur in spring and autumn, and if you're standing on the equator at noontime the Sun passes directly overhead and your shadow disappears beneath your feet. The morning of either equinox is easy to spot anywhere in the world, as the Sun rises due east (at right angles to the celestial pole you observed). It is this standard equinoctal hour (which can be captured from a sundial by an hourglass for comparison later) that mechanical clocks are set to count. Sundials display what is known as apparent solar time, which can deviate by as much as 16 minutes from the mean solar time kept by mechanical clocks with their fixed equinoctial hour. With the proliferation of mechanical clocks came a potential source of confusion, however— which of the two time systems do you mean: the uniform hour of machinery, or solar time, counting the number of hours since sunrise? Thus from the fourteenth century on it became necessary to specify a time as an hour "of the clock," such as "three o'clock."

Indeed, there's an even deeper historical link between the modern

---

4 All clocks are in essence devices for counting the oscillations of some regular process and displaying the tally. Modern clocks are no different in principle, they just exploit different physical phenomena with more rapid and more precisely regular ticks: counting the electronic oscillations of a quartz crystal in a digital watch or the microwave oscillations of a cloud of cesium in an atomic clock.

5 Although the day will actually be slightly longer because the sunlight is deflected in the Earth's atmosphere to give a period of twilight in the morning and evening.

clock face hanging on your wall and ancient sundial technology. Mechanical clocks displaying the time by an hour hand twirling around a dial were designed to be intuitively understood by people accustomed to reading the shadow line of a sundial. They first appeared in medieval European cities, and in the northern hemisphere the shadow from a sundial gnomon always rotates the same way: the direction we therefore adopted as "clockwise" for the hour hand. If during the reboot a mechanically advanced southern civilization reinvents the clock, the hands might instead turn in the direction we would consider anticlockwise.

So much for keeping time during the day. What can you do, working from the very basics, to track longer cycles of time—feeling the pulse of the seasons and reconstructing a calendar?

RECONSTRUCTING THE CALENDAR

Let's go back to our stick in the ground. We've already seen how you can follow the shortening and lengthening of its shadow during a day to find the time of noon. If you jot down the length of the noontime shadow on successive days, essentially measuring the maximum elevation angle of the Sun, you'll notice a periodicity over the seasons as the Earth orbits the Sun.[6]

If you stay up a bit later and monitor not the Sun's motion but the

---

6 How can you prove it is the Earth that moves around the Sun, and not vice versa (and thus that we are not in a privileged location in the center of the solar system)? All you need is a suitably accurate clock. Over a few nights you'll notice that any given star rises almost exactly four minutes later each night. If the only motion involved were the Earth turning on its axis like a top, the stars would wheel into view at exactly the same moment every night. But in fact the Earth's position shifts slightly, and so it takes a short while for its rotation to bring the same view as last night back into sight. Four minutes is $1/_{365}$ of 24 hours: the Earth has moved forward one day in its yearlong circuit of the sun.

nighttime sky, you'll have access to a much greater selection of celestial landmarks for subdividing the year and tracing your progress through the seasonal cycles. Many of the constellations visible from any particular location change during the year. For example, the familiar constellation of Orion the Hunter lies draped across the celestial equator, and so can be seen only in the northern hemisphere during the winter months. More exactly, individual stars are first visible and then disappear again on particular dates (allowing you to accurately count the 365 days in the year). These stellar events can be linked to the special days during the year that you've determined—the solstices and equinoxes—and so can be used to follow your progression through the year and anticipate the changing of the seasons. The ancient Egyptians, for instance, predicted the flooding of the Nile and the rejuvenation of their soils by the first appearance of Sirius, brightest star in the sky, which in our modern calendar equates to around June 28.

Thus, by noting down a few rudimentary observations you can reconstruct a year of 365 days[7] and pencil into the calendar the equinoxes and solstices that serve as four evenly spaced landmarks in the year—temporal monuments for the transition of the seasons and the coordination of your agriculture. Autumnal and vernal equinoxes—which as we've seen also serve to define your clock hour—fall around the 22nd of September and the 20th of March, respectively (for the north), and the solstices near the 21st of December and the 21st of June. So even if

---

7 In fact, you'll notice from your records kept over the first few decades of the reboot that with a 365-day calendar the date of the stellar events drifts steadily later in the year. This tells you that the year length isn't actually 365 days on the nose, but a fraction longer. (Thinking about it, there's no reason to expect the planet's orbit around the Sun to be an exact multiple of the time it takes to spin on its own axis.) Over 1,460 years a marker will slide backward through the whole year to arrive back on the original day you observed it. So relative to the celestial backdrop, the Earth spins an extra 365 days over 1,460 years. Thus there is an additional quarter day in each year that you need to take account of; otherwise your calendar will slip embarrassingly out of sync with the seasons. This is why in 46 BC Julius Caesar decreed the realignment of the date and introduced the leap year to ensure the calendar remain in step with the Earth.

the survivors regress so far after the apocalypse that the thread of history is severed by a period when no one keeps records, you'll still be able to work out the date by keeping your eyes on the celestial clockwork for a bit. If you wanted to, you could resurrect the Gregorian calendar, with its comfortably familiar structure of twelve months from January to December, and peg it back onto these special days you've determined.

But would it be possible to calculate what the year is after perhaps generations of no one ticking off the calendar? How long did the Dark Ages persist after the catastrophic subsidence of our civilization? One good way to find out relies on an astounding realization about the stars sprinkled across our night skies.

Over the course of a night, stars move around the sky like a vast dome with pinprick holes pirouetting above your head, each point of light maintaining a set configuration relative to others: the patterns of the constellations. The mind-blowing reality, however, is that over timescales immensely longer than a human lifespan, all the stars are actually moving past one another. If you were to fast-forward time again (this time counteracting the wheeling from Earth's spin), you could watch the stars sliding among one another, swirling across the sky like flecks of foam on a dark ocean. This is known as proper motion, and is due to the other suns whirling around the galactic center on their own orbital trajectories.

The particular target of most interest for determining the year at an unknown point in the near future is known as Barnard's star. This is one of the stars closest to the Earth, but it is a tiny, ancient sun, glowing with only a pitiful dim red gleam, and so despite its neighborly proximity still cannot be seen with the naked eye. Barnard's star can be easily picked out with a modest telescope, with a lens or mirror only a few inches across, though. Despite the slight trickiness in observing this star, it can serve as a natural time marker in the heavens. Barnard's star, due to its nearness, has the fastest proper motion of any known

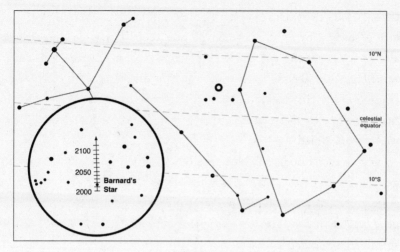

BARNARD'S STAR HAS THE FASTEST PROPER MOTION ACROSS THE
NIGHT SKY, AND OBSERVATIONS COULD BE USED TO REESTABLISH
THE CURRENT YEAR AFTER A BREAK IN THE HISTORICAL RECORD.

star in the heavens. It tears across the sky at almost three one-thousandths of a degree every year. This may not sound like much, but compared to all the surrounding stars it's blistering, and over a human lifespan it races almost half the diameter of a full Moon. So to find the date in the future, all a recovering civilization would need to do is make observations—even easier using photography—of the patch of the sky shown in the figure above, note the present location of Bernard's star, and read off the current year from the timeline.

Over a much longer timescale you can take advantage of the axial precession of the Earth. Like a spinning top, our planet's axis of rotation topples around in a circle gradually over time. The North Star, Polaris, happens to sit in line with the current orientation of the Earth's spin axis, and so is the only point that doesn't seem to wheel around the heavens. Right now, there's no equivalent South Star, as Earth's axis is currently passing through a barren region of the southern sky. Within a millennium the north pole will have wandered through

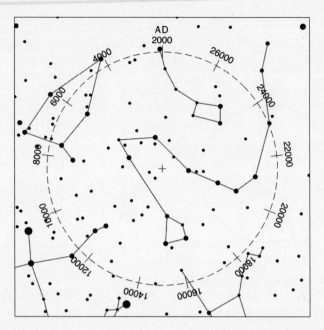

THE CIRCLING OF THE CELESTIAL NORTH (TOP) AND SOUTH (BOTTOM) POLES
AS THE EARTH'S SPIN AXIS PRECESSES OVER THE NEXT 26,000 YEARS.

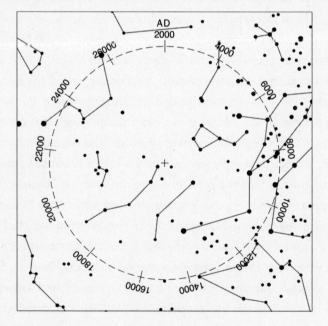

blank sky to pass close to other stars, and by 25,700 AD it will have roamed in a complete circle to return to its position at the birth of Christ. (Another consequence of this roving is that the points where the Sun's path crosses the celestial equator, the vernal and autumnal equinoxes, drift across the sky, and so this process is also called precession of the equinoxes.) It's a relatively straightforward task to watch where the contemporary celestial pole lies, especially if you've redeveloped basic photography and can image the star trails smeared out by Earth's rotation (with an exposure of a quarter hour or so). Compare it to the star map timelines shown on page 264 to read off your current millennium.

Recording the different motions of the Earth will allow you to tell the time of day and reconstruct a calendar to anticipate the changing seasons for agriculture. But how do you go about determining where you are on the Earth, and by extension, how can you learn to navigate your way effectively between different locations?

WHERE AM I?

Roving across terrain between familiar landmarks, or following the coastline by ship, is easy enough. But away from these comforting guides—crossing the featureless expanse of the ocean, for example— what can you do to make sure you're heading in the right direction? Chinese sailors first employed the incredible direction-seeking behavior of natural lodestones (in Middle English meaning "leading stone") in the eleventh century, and later magnetized iron needles. The compass needle works by turning itself to lie parallel with the lines of the Earth's magnetic field, and so aligning its length between the poles: it helps to mark the north-pointing end of the needle. Not only will a compass enable you to maintain a constant heading in total absence of any other external references, but if two (or more) prominent landmarks are in sight, you can take a compass bearing to them and so triangulate your position accurately on a map or chart. Although you can always find north or south by a clear night sky, the compass is a fantastic navigational tool when it's overcast. Beware, though: the celestial pole, caused by the Earth's rotation, and the magnetic pole, caused by the Earth's churning iron-rich core, do not perfectly match with each other. The discrepancy is only a few degrees at the equator, but as you voyage toward either pole, the compass error toward true north worsens.

If you're forced to return to absolute basics and can't find any relic magnets, you can always create a temporary magnetic field using electricity. We saw in Chapter 8 how a rudimentary battery can be built from an alternating stack of two different metals, which will push current along a lump of copper drawn into a wire and wrapped into a coil to form an electromagnet. While energized, this can be used to permanently magnetize an iron object, such as a thin needle suitable for a compass. (And if you're really starting from scratch, check Chapter 6 for how to smelt the metals in the first place.)

A compass will tell you a direction, and, combined with a previously mapped chart and landmarks, can also give your location. But what about a more general system for determining your position at any point on the surface of the Earth? It turns out that the solutions to the two fundamental problems addressed in this chapter—what time is it, and where am I?—are actually more deeply connected than you might have realized.

The first issue to be resolved in determining your location is to devise a system of unique addresses for all points around the planet. It's fine to describe the lake as being three miles out of town to the southwest, but what about locating a newly discovered island, or, indeed, your present position in the middle of the featureless ocean? The trick is to find a natural coordinate system for the globe of the Earth.

Finding your way around a city like New York with a regimented grid layout is relatively easy. The avenues all run roughly northeast with the streets cutting across them at right angles, and most of both are sequentially numbered. Getting anywhere in Manhattan is trivial: you walk up the current avenue until you reach the intersection with the street you want, and then along that street until you hit the destination. The address of a venue in midtown Manhattan can be as simple as listing the intersection it sits on: 23rd Street and 7th Avenue. Or if everyone agreed to a convention of always saying the street number before the avenue, all you would need is a pair: 23, 7 or 4, Broadway. An address here is far more than a label: it is a pair of coordinates that precisely pinpoints the location within the city. And by looking up at the signs on an intersection to see your own current position within the grid, you can instantly work out the direct course to steer to your destination, working along and across the blocks.

A similar coordinate system works for the entire planet. The Earth is an almost perfect sphere, with the axis of its rotation defining a north and south pole, and the equator as the circular line running around the planet's midriff. Due to this spherical geometry, it makes sense to divide up the area not with lines spaced at regular distances, as an ideal city grid might be, but with lines spaced at regular angles. So imagine standing at the north pole and shooting a line due south all the way around the planet to the south pole, and then swiveling around 10 degrees and shooting another line, and another, until you've turned around through 360 degrees of a complete circle. Similarly, you can start at the equator, already defined as the circle around the planet halfway between both poles, and imagine dropping shrinking rings every 10 degrees toward the north and south, with the poles therefore at 90 degrees.

These traces running north-south between the poles are called lines of longitude, and the east-west rings encircling the planet on either side of the equator are lines of latitude. All the lines of latitude are parallel to one another, and the lines of longitude cut across them always at right angles. So, near the waistband of the world the latitude-longitude coordinates approximate the street-avenue system on the flat plane of Manhattan, with the square grid increasingly distorted toward the poles by the spherical geometry of the Earth. As with the Manhattan streets, you need to set starting points so you can specify your numbered coordinates relative to them. The equator is the obvious 0° latitude line, but there is no corresponding natural zero mark for the longitude numbering: we happen to use Greenwich, London, as the "prime meridian" purely out of historical convention.

To specify your location anywhere on the planet using this universal address system, all you need to do is state how many degrees north or south of the equator you are—your latitude—and how many degrees east or west of the prime meridian—your longitude. Right now, my smartphone says that I am at 51.56° N, 0.09° W (I'm in north London, not far from Greenwich).

So, the original problem we set ourselves—how to navigate the world between known locations—splits neatly into two separate questions: How do I find my latitude, and how do I find my longitude?

Latitude is actually pretty easy to establish—the richly patterned night sky offers more than enough information. Polaris, the stationary bull's-eye in the circling star trails, hangs directly above the north pole, so it stands to reason that your angular distance from the equator is the same as the angle between this celestial pole and the horizon. The problem of determining your latitude on the Earth directly translates to measuring the elevation of stars.

Most simply, you could build a navigational quadrant from odds and ends lying around. A quarter circle of cardboard or thin wood is marked with a subdivided angular scale between 0° and 90° around the curved arc. Two notches are put on the ends of one of the straight edges so that a target can be sighted along it, and a plumb line fixed to the corner dangles straight down to indicate the angle of elevation against the scale. Although not particularly sophisticated, such a basic device will still allow you to sight the polestar and thus discover your latitude on the Earth, to an accuracy of several degrees, which equates to finding how far north of the equator you are to within a few hundred kilometers.

A far more elegant and accurate instrument was developed in the 1750s, and is still widely used today as a backup navigational device in case of power failure or loss of GPS. The sextant is based on a sector one-sixth of a full circle (and is named for this attribute, fulfilling the pattern set by the earlier quadrant, then octant), and can measure the angle between any two objects. Most usefully for navigation, the sextant can very precisely give you the elevation angle above the horizon of the Sun, or Polaris, or indeed any other star. The design of such a marvelously useful contraption is easy to replicate in retrospect, and as soon as your rebooting civilization has recovered the basic capabilities for shaping metal, grinding lenses, and silvering mirrors, you've nailed the prerequisite technologies for the sextant.

The frame of the sextant is a 60-degree wedge of a circle, much like a slice of pizza held vertically, the tip toward the sky. A rotatable arm is pivoted at the tip and hangs down to point at an angle scale running around the curved rim. The key component of the sextant is a half-silvered mirror mounted on the front edge, so that the operator can still look forward through it. An angled mirror on the pivot of the arm reflects the image of whatever it is pointed at down to the half-silvered mirror, and so superimposes the two views for the operator.

THE SEXTANT, WITH ITS SIGHTING TELESCOPE (A), HALF-SILVERED MIRROR (D), AND ANGULAR SCALE (H).

To use your sextant, look through the small sighting telescope and tilt the whole instrument to line up on the horizon through the forward half-mirror. Now rotate the arm until the reflected apparition of the Sun, or any target star, swings down and seems to perch right on the horizon (some darkened pieces of glass can be inserted between the mirrors to safely reduce the glare). The angle of elevation is indicated on the bottom scale by the swing arm.

Once you've relearned the patterns of the heavens and recorded tables of the positions of the brightest, landmark stars for different dates and times, you can then take a sighting off any of these to determine your latitude even if the polestar is obscured. And once you've tabulated the noontime height of the Sun for different dates and latitudes, you can use a sextant and calendar to backtrack and find your latitude while you're on the move during daytime, too. Once you know how to read it, the sky is a fantastic combination tool—both compass and clock for local time.

The second half of the coordinates needed to pinpoint where you are, the longitude, is unfortunately not nearly as tractable. It's hard to use the heavens for finding how far east of the prime meridian you are, because the rotation of the Earth is constantly rolling you around in that direction. To extend the New York analogy to the breaking point, seventeenth-century sailors could easily tell which street they were on, but working out the avenue was next to impossible. Their only recourse was to sail by dead reckoning—extrapolating their bearing and estimated speed, and hoping they weren't being pushed too much off course by unknown currents—to the right latitude at a point they could be confident hadn't overshot the target, and then sail due east or west along that latitude until hopefully they stumbled into their destination.

The Earth spins toward the east, causing the apparent motion of the Sun across the sky and the wheeling of the stars at night. The Sun's position is how we define the time of day (right back to the fundamentals of sundials we saw earlier), and so the problem of finding your longitude—how far around the world you are from some chosen baseline—boils down to finding the difference in time of day at the same instant between the baseline and your current position. The Earth spins 360 degrees in 24 hours, so a difference in the timing of noon of one hour equates to 15 degrees of longitude. Determining your longitude is therefore a measure of time transposed back into space. In fact, you've almost certainly felt the solution to longitude acutely yourself: modern high-speed air transport teleports us between remote locations with very different local times before our bodies can adapt. Before GPS, navigators exploited the same principle as that behind jet lag!

So to find the vital second coordinate to pinpoint your position, you can use a sextant to find the local time where you are now, and compare it to the current time back at the prime meridian. The trouble, though, is how to communicate that baseline time to remote regions of the globe.

The longitude problem was finally cracked in our history by the invention of suitable clocks: impervious to the pitching and rolling of a ship on the high seas and sufficiently accurate over the months to years of a voyage. Clearly a pendulum and drive-weight system would be useless for a marine clock; it is the spring that provides the solution to both of these functions. A suitable oscillator can be made from a balance spring: a thin strip of metal coiled into a spiral around the shaft of a weighted ring that springs back and forth. The function is similar to that of a pendulum, but with the restoring force at the extremities of the oscillation generated by the tightening of a spiral spring instead of gravity. A spiral spring, wound tightly to store energy in its tension, can also provide the motive force to drive the clockwork. This is a far more compact power source than a steadily descending weight, but employing a spring for this introduces a new problem that must itself be solved with another invention. The trouble is that the force exerted by a spring changes as it unwinds: strongest at first and then progressively weaker as its pent-up tension is released. The best method to even out this power, and so regulate the rate of the clock, is to connect the free end of the coiled spring to a chain wrapped around a cone-shaped barrel known as a fusee. This way, as the spring unwinds it acts further and further up the fatter end of the fusee and so benefits from an enhanced leverage effect that neatly compensates for its reduction in force.

A suitably complex clock, incorporating mechanisms to automatically compensate for swings in humidity and temperature (which affect the thickness of lubricating oils and stiffness of springs) and other sources of variation, is a miraculous device, an almost magical cage that can hold time itself, perfectly preserved, like a trapped genie.[8] The trouble, for trying to leapfrog straight to this point during the rebuilding of civilization, is that even knowing the solution to a problem is sometimes not enough. The devil is often in the exquisitely refined detail, and there may not always be shortcuts or opportunities for leapfrogging during the recovery of advanced civilization. It took the monomaniacal, obsessive efforts of a single clockmaker, John Harrison, over much of his life, to design and construct a sufficiently accurate marine clock, and necessitated the invention of many new mechanisms along the way, including caged roller bearings to greatly reduce friction and the bimetallic strip to cancel out expansion with temperature.

So is there a different way around the problem? Obviously, if any reliable clocks or digital watches have survived in the aftermath, all you would need to do is set one to the local time when you head off,

---

8 Large surveying ships would often carry several chronometers, to average out errors and for multiple redundancy. HMS *Beagle* set out in 1831 with no fewer than 22 chronometers to ensure an accurate fix on the location of foreign lands (including the Galápagos Islands, where Darwin's observations of the wildlife set him toward his revolutionary theory).

pop it in your pocket during your odyssey, and take it out to compare against local time (which you'd still need to determine by sextant observation) to get a fix on your current longitude. But what if there are no relic timepieces?

The problem in the early eighteenth century was that although you could work out your local time, you had no way of remotely telling the current time back in Greenwich. Harrison's eventual solution was to take a copy of Greenwich time with you, but it would equally work if Greenwich could somehow regularly communicate its time to ships around the world. One harebrained proposal was to set up a network of signal ships anchored in mid-ocean to relay the sound of cannon blasts indicating the moment of noon in London. But we now know of a far more practical means for signaling over vast distances: radio.

A post-apocalyptic civilization rebooting along a different pathway through the web of scientific discoveries and technologies may arrive at another solution to global navigation. They might find that building rudimentary radio sets (see Chapter 10) is an easier prospect than the fabulously intricate workings and compensatory mechanisms of a sufficiently accurate timepiece. (But then, this will obviously depend on the rate of reattainment of different technologies—how do you compare the complexity of minute mechanical cogs and springs to that of electronic components?) Regular timing signals can be broadcast from whichever prime meridian is selected as the baseline for longitude, and relayed to far-flung regions by ground stations or other ships. In this way, the sight you could see in the early stages of recovery might be wooden-hulled sailing ships coursing across the world's oceans, looking very much as they did in the Age of Sail, but with a subtle difference: a metal wire strung up the main mast as a signaling aerial.

The bright urban illumination and light pollution of modern industrialized civilization has robbed many of us of our formerly intimate relationship with the facial features of the heavens. But in the aftermath you'll need to become reacquainted with the configuration of the sky in order to reclaim your connection with the rhythm of the seasons. This isn't irrelevant, arcane astronomical detail: it will enable you to plan the agricultural cycle to avoid starving to death, and it will prevent you from getting lost in the wilderness.

CHAPTER 13

THE GREATEST INVENTION

We shall not cease from exploration
And the end of all our exploring
Will be to arrive where we started
And know the place for the first time.

T. S. ELIOT, *Four Quartets* (1943)

THROUGH THE PAGES of this book we've covered many topics that are utterly critical to any civilization, such as sustainable agriculture and construction materials, as well as some advanced technologies that will be required once a recovering post-apocalyptic society has progressed to a more developed stage generations after the Fall. We've explored shortcuts through the web of knowledge, what gateway technologies to aim for, and how to leapfrog over intermediate stages to superior, yet still achievable, solutions.

But despite all the vital knowledge presented in this reboot manual for civilization, there's no certainty that a post-apocalyptic society will attain an advanced technological state. Many great societies have flourished thoughout history, their wealth of knowledge and technological prowess a glittering gem in the world at the time, but most stall at some point and reach a stasis, an equilibrium state with no further progression, or they collapse altogether. In fact, the sustained progress of our current civilization is something of a historical anomaly. Medieval European society progressed through the Renaissance, the agricultural and scientific revolutions, the Enlightenment, and finally the Industrial Revolution to create the mechanized, electrified, globally interconnected society we live in today. But there is nothing inevitable about a sustained trajectory of scientific development or technological innovation, and even vibrant societies can lose the impetus to advance further.

China provides a particularly interesting case in point. For many centuries the Chinese civilization was technologically vastly superior to the rest of the world. China had invented the modern horse collar, the wheelbarrow, paper, block printing, the navigational compass, and gunpowder—all world-changing inventions that we've covered throughout this book. Chinese textile production made yarn using multiple spinning frames with a centralized power source, and operated mechanical cotton gins and sophisticated looms. The Chinese mined coal, discovered how to convert it into coke, utilized large vertical waterwheels and trip-hammers, and beat Europe by one and a half millennia to using blast furnaces to produce cast iron, then refining it into wrought iron. By the end of the fourteenth century, China had achieved a technological capability not seen anywhere in Europe until the 1700s, and seemed poised to initiate an industrial revolution of its own.

But, astonishingly, as Europe begun to emerge from its long Dark Ages with the Renaissance, Chinese progress faltered and then ground to a halt. China's economy continued to grow, mostly due to internal trade, and the expanding population enjoyed consistent living standards. But no further significant technological advancement occurred, and, indeed, some innovations were subsequently lost again. Three and a half centuries later, Europe had caught up, and Britain plunged into the Industrial Revolution.

So what was it about eighteenth-century Britain, and not fourteenth-century China or indeed another nation in Europe at the time, that fostered this transformative process—why *there* and why *then*?

The Industrial Revolution encompassed increases in the efficiency of textiles production—the mechanization of spinning and weaving, transplanting these traditionally small-scale, home-based activities into large, centralized cotton mills—as well as advances in iron making and steam power. And once industrialization got underway, the process fed itself, and the transformation accelerated: coal-fired steam engines pumping out mines allowed more coal to be extracted, which fueled blast furnaces to produce more iron and steel, which in turn were used to build more steam engines and other machinery. But the conditions that made all this possible in the first place were quite specific. While a certain proficiency in engineering and metallurgy was of course required to construct machinery for alleviating the toil of humankind, the key trigger for the Industrial Revolution wasn't knowledge. It was a particular socioeconomic environment.

There has to be some benefit to building a complex and therefore expensive piece of machinery or a factory to accomplish what is already being achieved by people using traditional methods. And eighteenth-century Britain presented a peculiar confluence of factors that provided the necessary impetus and opportunity for industrialization. At that time Britain possessed not only abundant energy (coal), but an economy with expensive labor (high wages) coupled with cheap capital (the ability to borrow money to undertake large projects). Such circumstances encouraged the substitution of capital and energy for labor—replacing workers with mechanization such as automated spinners and looms. The economic situation in Britain had the potential to generate enormous profits for the first industrialists, and it was this that provided the incentive to put up large amounts of capital to invest in machinery. On the other hand, China at the end of the fourteenth century, despite coal mining, coke-fired blast furnaces, and mechanized textiles manufacture, did not have conducive economic conditions in place to drive an industrial revolution. Labor was cheap, and would-be industrialists could expect little benefit from innovations that improved efficiency.

So while scientific knowledge and technological capability are necessary prerequisites for the advance of civilization, they are not always sufficient. If a post-apocalyptic society is knocked back to a rudimentary pastoral existence, there is no guarantee it will eventually undergo an Industrial Revolution 2.0, even with all the crucial knowledge provided in this book. In the end, social and economic factors determine whether scientific investigation flourishes or innovations are adopted. Throughout this book there has been the underlying assumption that the survivors in a post-apocalyptic civilization would want to progress along our developmental trajectory to an industrialized life. While I don't want to get into a debate over whether technology necessarily makes people happier, it is a robust point, I think, that a community struggling for subsistence, with an uncomfortable and punishingly hard lifestyle and access to only basic healthcare, would certainly appreciate the application of scientific principles to improve their standard of living. But at what point does a technologically progressing civilization reach a peak beyond which further advance brings diminishing returns? Perhaps a recovering civilization will reach equilibrium at a certain technological level, neither advancing further nor regressing, once it has achieved a stable economy, comfortable population size, and the ability to draw sustainably on natural resources.

THE SCIENTIFIC METHOD

This book is of course not a complete compendium of all the information you would need to rebuild your world from scratch. A great deal of material has necessarily been left out. We've mostly focused on inorganic chemistry, useful for making agricultural fertilizers or industrial reagents, rather than the synthesis or transformations of organic molecules. Organic chemistry has become increasingly important over the past century: processing the fractions of crude oil, purifying and modifying natural pharmaceutical compounds into more potent versions, synthesizing pesticides and herbicides for more reliable food production, and creating a whole new domain of materials with properties unlike anything we find in nature: plastics.

We've talked about biology to the extent of how you can nurture certain animal or plant species, or control microorganisms, in order to feed yourself and remain healthy. But we've not looked at the details of how life actually works on a molecular level—why it is, for example, that we need to breathe in oxygen and exhale carbon dioxide, whereas plants drive the opposite chemical process using the energy of sunlight.

We've skipped a lot of materials science and engineering principles and only brushed over the building blocks of all stuff: the structure of the atom and the four fundamental forces of nature. Not all atoms are stable, and radioactivity offers the possibility of an appallingly destructive weapon, as well as a source of peaceful power, but also allows you to determine the age of our planet, offering a glimpse down the dizzying hole of deep time. In Earth sciences we've missed out on the theory of plate tectonics, for example: the mind-blowing concept that the vast continents are scudding across the surface of the planet like leaves on a windy pond, occasionally crunching into one another to crumple up entire mountain ranges. These profound realizations that the world has not always been as it is now, and is bewilderingly old, are required to understand the theory of evolution by small changes from one generation to the next. All of these represent kernels of knowledge that a recovering society would need to reexplore and unpack for themselves by investigation, as well as by filling in the gaps between the other hints provided in this book, before eventually reconstituting the cornucopia of knowledge we collectively hold between all of us alive today.[1]

So how do you find things out for yourself? What are the tools you need to relearn the world? Let's continue with our back-to-basics approach from the previous chapter and look at the most effective strategy for producing new knowledge yourself: science.

The basis behind all scientific investigation is the appreciation that the universe is essentially mechanical, its components interacting with one another in orderly ways following universal governing laws and not the whims of temperamental gods. These underlying rules can be revealed by reasoned thought based on firsthand experience and observation. First and foremost, science is empirical, and everything must, in principle, be checked and verified independently, rather than basing it on logic alone or merely accepting the proclamations of past or present authorities (or, indeed, this book you're holding in your hands). So if you want to manipulate the world around you for your own benefit, to create artifacts or pieces of technology that exploit particular effects, you must first develop a sound comprehension of the natural laws that the world abides by. This understanding can come only from observing the world and spotting patterns in its behavior. But just as important, you need to have the capacity to notice discrepancies in the expected pattern: anomalies that betray new natural phenomena—the compass needle twitching next to a wire or the halo around a mold patch cleared of bacteria, for example. This requires the ability to measure things accu-

---

1 I'm sure that many people reading this book have been surprised that a particular topic they consider important has been overlooked. As far as possible within the bounds of a single book, I've tried to include as much as I think would be absolutely indispensable knowledge for rebooting. You could rebuild a functioning technological civilization without knowledge of human evolution or the planets of the solar system, but not without being able to effectively maintain the fertility of your fields or produce chemical alkalis. I'd be keen to hear from you via this book's website, the-knowledge.org, as to what you personally would consider vital knowledge for accelerating the rebuilding of civilization from scratch, and why.

rately, to be able to place numbers or values on different aspects of nature to compare them and monitor how they change over time.

The absolute root of science, then, is the careful design and construction of instruments for making measurements, as well as units to count these in. For example, a straight stick marked with regular notches is the simplest kind of instrument: a ruler for measuring length. But in order to communicate to someone the size of an object you've measured as 6 notches long, they also need to know the unit you are using—the exact spacing between the notches. Hence the key to recovering science from scratch lies in the creation of a set of measuring units. A post-cataclysmic society will need a system of measures in any case. The basic functions of civilization include the marking of distances for construction or travel, the measuring of fluids in a jug or weighing of solid produce for trade, the administration or taxation of areas of agricultural land, and the timing of different civic activities during the day. We experience these fundamental properties—length, volume, weight, and time—directly with our senses, and they are easily quantified. Other properties, such as heat or the tingle of electric current, we also encounter with our senses, but we need cleverly designed instruments to be able to measure them.

THE TOOLS OF SCIENCE

Most societies devise their own system of measures for distance, volume, or weight. These units are usually on a human scale relevant to everyday life: a pound weight represents a handful of meat or grain, and the second is a division of time corresponding roughly to the heartbeat. Indeed, many of these traditional units have been directly based on the dimensions of the body, such as the foot, inch (thumb), cubit (forearm), and mile (one thousand Roman paces). However, the problem with these units is that they vary not only from person to person, but often involve incredibly cumbersome conversion factors: the mile, for example, is equivalent to 1,760 yards, 5,280 feet, or 63,360 inches. What you ideally want is a standardized set of units that are interrelated and incorporate a convenient hierarchy of scale.

The system used today throughout the global scientific community, and almost universally for national administration and commerce, is the metric system devised in the 1790s amid the reorganizing fervor of the French Revolution.[2]

This international system of units (SI is the French acronym) defines just seven fundamental units, including those for length, mass, time, and temperature; every other measurement can be naturally derived from combinations of these units. Smaller or larger multiples of the core unit are restricted to the convenience of base ten, and indicated with an agreed prefix. For example, the meter is the standard unit of length, with smaller objects described in parts of a meter—a centimeter as a hundredth, a millimeter as a thousandth—and larger distances as multiples, such as a kilometer stretching 1,000 meters.

Alongside the meter, a second base unit is that of time—the second. Building from just these two base properties, using combinations or ratios of them, you can derive a great many other units. Multiplying two distances together (such as the length and width of a rectangular field) yields a measure of area, and consequently area always has units of distance squared. Multiplying three dimensions gives a volume, with units of length cubed. Dividing a quantity by time tells you how quickly it is varying—giving you a rate of change. So dividing a distance by time provides a unit of velocity, such as kilometers per hour, and divid-

---

2 About the only countries that haven't fully adopted this system are the US and the UK, where outdated units persist with the use of miles on road signs and car speedometers and drinks served in restaurants and pubs by the pint. The historical reason was Napoleon excluding the English-speaking world when he convened a congress in 1798 to encourage international adoption of the new metric system—the British had just sunk the French fleet at the Battle of Aboukir and so were not invited to the party.

ing by time again indicates how rapidly something is speeding up or slowing down: acceleration and deceleration. Units can be combined in ever-deeper degrees of derivation to describe further physical properties. The kilogram is the base unit for mass, and the density of a body—and so whether it will float or sink—is found by dividing its mass by volume. Combinations of mass and velocity yield measurements of the momentum and energy of a moving object.

So how can you reconstitute this system of measures and units from first principles in the post-apocalyptic world, if no graduated jug, set of scales, working clock, or thermometer can be found?

Starting with the meter as the primary base unit, you can derive many others from it. Build a cube-shaped container with each interior side exactly 10 centimeters long (one-tenth of your meter). The internal volume of this box is 1,000 cm^3, or one liter. Fill the container with ice-cold, distilled water, and the water will have a mass of exactly one kilogram. Use a set of well-constructed balance scales (hang a straight, stiff rod from its midpoint if you need to) and you can use this liter of water to create any fraction or multiple of this unit by moving the mass closer to or farther from the pivot. To bring time into the fold you can utilize the pendulum we encountered in the last chapter. The length of a pendulum that swings each way (i.e., a half-period) in exactly one second is 99.4 cm, and even if you used a meter-long pendulum it would still be accurate to within three milliseconds—a hundred times less than the blink of an eye.[3] So, building from the meter alone you can reconstruct the metric units for volume (liter), mass (kilogram), and time (second).

---

3 In fact, historically, the argument ran the other way, and it was proposed in the seventeenth century to define the meter as the length of a pendulum that has a half-period of exactly one second. This is why the word "meter" shares its meaning with the rhythm of poetry or music. The proposal was abandoned in favor of an alternative based on the dimensions of the Earth, however, due to the effect that variations in the local gravitational strength across the face of the planet have on a pendulum's rhythm.

But how do you define the length of the meter for the survivors of the apocalypse, to allow them to unpack everything else from it? Well, the line drawn along the bottom of this page is exactly 10 cm long, and so from this seed the other units can be reconstructed.

All of these quantities discussed so far can be measured with very rudimentary implements—a graduated ruler or balance scales—but how would you go about devising from scratch an accurate gauge, meter, or instrument for measuring less physically tangible attributes, such as pressure or temperature? The general principles needed to design novel instruments will be essential for the scientific scrutiny of the inner workings of the world, particularly when you stumble across strange new effects and want to understand them.

One of the very first scientific instruments you will need to invent is intimately related to the puzzling observation that a suction pump can never raise water from a well more than about 10 meters, as we saw in Chapter 8. Fill a long tube with water, seal it at both ends, and then dangle it down out of a tall tower. Dip the lower end into a basin of water and remove the bottom seal. Water will flow down out of the tube by gravity, but not all of it, and you'll find that no matter how you set up the experiment, the remaining water column is always about 10.5 meters tall (curiously, this is the same as the maximum height that a suction pump can raise water from a well). At the top of the tube you'll notice a clear space left behind as the water drained away and where the air has not been able to reenter—a vacuum. The weight of the water column is held up by the force exerted at the bottom by the overbearing ocean of air—the atmosphere. Changes in the surrounding pressure are revealed in the rising or falling height of the column: it is a working pressure gauge. Using a denser liquid makes for a more practical barometer, and atmospheric pressure equates to only 76 cm of mercury (rather than more than 10 meters of water).

Such a barometer can be built out of any glass tube—and the elegance of such an arrangement is that it is naturally invariant in relation to the diameter of the tube used (as long as the diameter is constant along its length). The thicker the mercury column the more weight is pulling it down, but this is perfectly balanced by the increased force of atmospheric pressure pushing it back up—any mercury-column barometer will immediately give you the same answer, regardless of the details of its construction.

Once a novel instrument becomes available, it offers an unprecedented means for investigating the world and often leads to a rapid burst of new discoveries. For example, try hiking your new barometer up a mountain to explore how atmospheric pressure changes with altitude, or look for patterns and correlations between the finely fluctuating air pressure at your location and the weather. Medics today still quote blood pressure in units of the height of a corresponding mercury column: around 80 mmHg is the normal value between heartbeats.

Measuring temperature demands a little more cunning. The temperature of different objects is revealed to us by our own senses—we can feel whether something is hot or cold. But how do you build a device to precisely measure that subjective experience, to put a number on something's hotness? The trick is to look for physical effects that correlate with your personal sensation: you'll notice that as substances get hotter they also often expand. The next step then is to build a device designed to exploit this physical phenomenon for an objective expression of temperature. A simple heat-sensing device can be constructed with a long thin tube of glass, partially filled with liquid, and then sealed at both ends—such an arrangement maximizes the visible effect of expansion. Strap the tube to a ruler, and the height of the top of the fluid column provides a proxy for the temperature encountered. You can now measure the temperature of objects relative to each other, independent of your subjective perception.

But the fluid column height seen at different temperatures in a particular instrument, and thus the measurement you get, will be entirely dependent on the dimensions and other idiosyncrasies of its construction (unlike the simple barometer we already looked at): you won't be able to compare your results against anyone else's. What you need is a standardized scale that anyone can derive and mark on their own instrument. And for that you need a way of determining fixed points: events or states of matter that always occur at exactly the same temperature and so can serve as a thermometric benchmark. It seems natural to base a temperature scale on water, as the changes in state of this substance occur over a range relevant to everyday life—from an icy winter's morning to a steaming saucepan. Once you've got an upper and lower fixed point nailed down, it's then a simple matter of regularly subdividing the range in between into a convenient round number of graduations to give a meaningful temperature scale. The Celsius scale is based on the freezing and boiling of water[4] as fixed points, defined to occur at 0 and 100 degrees, respectively. But rather than using water itself as the fluid, you'll realize that mercury expands far more uniformly for an accurate thermometer. For thermometers capable of operating at temperatures beyond the boiling point of mercury, for use in a kiln or furnace, for example, you will need to exploit other physical phenomena. Your investigations of electricity, for instance, will reveal that the resistance of a wire often increases with temperature.

THE SCIENTIFIC METHOD—CONTINUED

This, then, is the fundamental process for devising reliable means for measurement of any attribute. As the recovering civilization discovers

---

4 In actual fact, since the process of boiling depends on other factors such as the roughness of the container for forming bubbles, the temperature of a saturated cloud of steam at atmospheric pressure is a more consistent and reliable standard.

strange new phenomena of nature, new fields of scientific research emerge. Means of isolating the properties of these phenomena and translating them into something that can be reliably measured must be devised before they can begin to be understood and exploited for technological applications. For example, when electricity was first stumbled upon, investigators struggled to quantify the properties of this new phenomenon, resorting to subjectively rating the intensity of the shock they received. But as the phenomenon was investigated, some of its repeatable effects were noticed and could then be employed for measurement—using the motor effect to deflect a needle around an ammeter dial, for example. And these scientific instruments aren't just gizmos for the laboratory: they are also the thermometer that reveals your child's fever, the meter monitoring the flow of electricity into your home, the seismometer serving as a sentinel for foreshocks presaging a larger earthquake, or the spectrometer detecting trace indicators in your hospital blood test.

These devices for measuring the world, and the standardized units they count in, are the basic tools of science. Knowledge of the world can be gleaned only by attentively inspecting it, or even better, by carefully arranging contrived circumstances to investigate a particular aspect in detail. This is the essence of the experiment.

An experiment is a way of artificially constraining a situation, to attempt to remove other distracting or complicating factors so you can focus tightly on how just a few features behave. An experiment is asking a clearly worded question of the universe and eagerly watching how it responds. Experimentation addresses the dissatisfaction with what nature happens to display for you, and forces it to reveal tightly defined facets of itself as you poke in different ways. Once you have controlled all the complicating factors and pinned down just one, you'll then move on to the next, and so on, systematically interrogating the system until you understand how all the parts fit together.

As well as your instruments to extend the human senses and to measure the results of different tests—a thermometer, a microscope, or a magnetometer—the meticulously constrained scenario demanded by a particular experiment often requires new contraptions: specially constructed scientific equipment designed to create specific conditions for you to study. Just as important, observations and results of your experiments need to be recorded numerically—adorning qualitative descriptions of what happened with measured, quantitative precision. But far beyond merely using enumeration to accurately compare different outcomes, the language of mathematics can be adopted as a powerful tool for precisely describing the behavior and patterns of nature, and the interrelationships between her parts. An equation summarizes a complex reality into its condensed essence. The upshot is that you can calculate the expected outcome in new, unobserved situations—you can make precise predictions.[5]

But for all of its careful observations, intricate experiments, and condensed equations, the absolute essence of science is that it offers a mechanism for you to decide which explanation is most likely to be the right one. Anyone with imagination can construct a tale that neatly accounts for the ways of the world—where rain comes from, what happens when something burns, or how the leopard got its spots. But these are no more than entertaining diversions—etiological Just So stories—unless you have a reliable way of selecting which one is more likely to be correct.

Scientists construct a best-guess story based on their prior knowledge and what's already been established, called a hypothesis, and design particular experiments targeted to test different predictions of this story—systematically poking and prodding the hypothesis to check

---

5 Mathematics is one topic that has not been covered in depth here. Calculations are clearly important for engineering designs, and mathematics is the language for the statement of physical law, but it does not lend itself to explanations of general principles within the scope of this book.

how well it works, or to inform the choice between competing proposals. And if this account withstands the tests of experiments or observations many times, and is not found wanting, then it becomes a well-founded theory and we can have confidence in using it to explain other unknown aspects. But even then, no theory is ever inviolate: it could itself be torn down later, undermined perhaps by new observations that it cannot account for, and replaced by an explanation that offers a better fit to the data. The essence of science lies in repeatedly admitting you were wrong and accepting a new, more inclusive model, and so, unlike other belief systems, the practice of science ensures that our stories become steadily more accurate over time.

In this way, science isn't listing *what you know*: it's about *how you can come to know*. It's not a product but a process, a never-ending conversation rebounding back and forth between observation and theory, the most effective way of deciding which explanations are right and which are wrong. This is what makes science such a useful system for understanding the workings of the world—a powerful knowledge-generating machine. And this is why it is the scientific method itself that is the greatest invention of all.

But in the hardships of a post-apocalyptic world, you'll not be immediately concerned with accruing knowledge for its own sake—you're going to want to apply that understanding to helping improve your situation.

SCIENCE AND TECHNOLOGY

The practical application of scientific understanding is the basis of technology. The operating principle of any technology exploits a particular natural phenomenon. Clocks, for example, utilize the discovery that a pendulum of a particular length always swings with the same rhythm, and this reliable regularity can be used to meter time. The incandescent light bulb capitalizes on the fact that electrical resistance causes wires to get hot, and that very hot objects emit light. In fact, anything but the simplest technology exploits a whole collection of different phenomena, controlling and orchestrating the various effects to achieve a designed purpose. New technology invariably builds on older ones, borrowing previously developed solutions for providing particular functions, like off-the-shelf components. It is often only the ingenious new combination of established parts that is novel in an invention, and we've looked closely at two examples: the printing press and the internal combustion engine. Each new technology offers a novel function or advantage, which can in turn itself be incorporated into further innovation—tech begets more tech.

As we have seen throughout this book, history has witnessed the intimate interaction of science and technology. Researchers discover an unknown phenomenon, principally by demonstrating that an observation cannot be explained by any known phenomena, and then explore its different effects and learn how to maximize and control them. Harnessing these extra principles allows the creation of tools or other inventions to ease human toil or enrich everyday life—the process of turning oddity into commodity. Exploiting novel principles also allows the building of new scientific instruments and experiments, to scrutinize and measure nature in fresh ways, and so drive yet more fundamental discovery and the unearthing of further natural phenomena. Science and technology are in a close symbiotic relationship—scientific discovery drives technological advance, which in turn enables further knowledge-generation.

Not all innovations draw directly on recent discoveries, of course— the spinning wheel is a product of pragmatic problem solving—and even the celebrated poster boy of the Industrial Revolution, the steam engine, was initially developed predominantly through empirical know-how and the practical intuition of the engineers rather than theoretical considerations. And indeed, there are examples in our history when inventors didn't correctly understand the operating principle behind their creation, but it worked nonetheless. The practice of canning food for preservation, for example, was developed long before the acceptance of germ theory and the discovery of spoilage by microorganisms.

Even with the correct scientific understanding of the phenomena involved, producing a working invention demands far more than a single leap of imaginative creativity. Any successful innovation requires a long gestation period of tinkering and debugging the design before it works reliably enough to be adopted widely—this is the 99 percent perspiration that the American inventor Thomas Edison described as following the 1 percent inspiration. The same process of rigorous, methodical investigation that drives science is applied here as well, analyzing in this case not the natural world but our own artificial constructs—experimenting with nascent technology to understand its shortcomings and improve its effectiveness.

Survivors of the Fall will appreciate the importance of scientific understanding and critical analysis, which will be crucial in maintaining relic technology as long as possible. But over the generations the post-apocalyptic society must protect itself against slipping into a rationality coma of superstition and magic, and must nurture an inquisitive, analytical, evidence-based mind-set for the rapid attainment of their own technological capability. This is the flame that the survivors must keep burning. It is by thinking rationally that we have been able to vastly improve our productivity in growing food, to master materials beyond sticks and flint, to harness power sources beyond our own muscles, and to build transportation to convey us much farther than our own feet ever could. It is science that built our modern world, and it is science that will be needed to rebuild again.

THIS BOOK CAN OFFER only glimpses of the vast architecture of current understanding and technology. But the areas we explored will be the most critical for nurturing a nascent culture through an accelerated reboot and enabling it to relearn all else. My hope is that by seeing just how civilization actually gathers and makes all the fundamentals we need, you'll come to appreciate, just as I have during the research for this book, the things we take for granted in modern life: bountiful and varied food, spectacularly effective medicines, effortless and comfortable travel, and abundant energy.

Homo sapiens first had a marked effect on the planet around ten thousand years ago, with the sudden disappearance of around half of the world's large mammal species—we are the prime suspect for driving this extinction with our teamwork and improved hunting technology of stone axes and tipped spears. Over the next ten thousand years there was a steady deforestation around the Mediterranean Sea and northern Europe as people settled and cleared the surrounding land. Three hundred years ago the human population began to grow rapidly, and gradually every scrap of land that was suitable for agriculture became cultivated. There were also profound changes not just to the landscape, but to the chemistry of the entire planet, as hundreds of millions of years of accumulated carbon was dug out of the ground and pumped into the atmosphere with mounting fervor. The rising carbon dioxide levels in the atmosphere pushed the very climate of the world, driving global warming, rising sea levels, and acidification of the oceans. Dotted towns and cities swelled and coalesced with each other like bacterial colonies, as roads were draped like ribbons across the rolling landscape, looped into rings around large urbanizations, and tangled up in gloriously complex overpasses at major interchanges. A growing swarm of metallic craft hurried back and forth over the land and seas of the world, crisscrossing the skies, and some even piercing out of the atmosphere. At night this ceaseless fervent activity was apparent from space, with the continents marked out in webs of artificial lights, networks of glowing nodes and lines.

And then silence.

The worldwide network of traffic abruptly halts, the web of light fades and dies, cities rust and crumble.

How long will it take to rebuild? How quickly can technological society recover after a global cataclysm? The keys to rebooting civilization may well be within this book.

THE-KNOWLEDGE.ORG

Explore the book's website for further material, recommendations, and videos, and continue the debate on the community page:

What knowledge would *you* preserve?

 @KnowledgeCiv

@lewis_dartnell
www.lewisdartnell.com

FURTHER READING
AND REFERENCES

A small selection of books discussing the historical development of science and technology have proved absolutely indispensable through many of the chapters of this book, and I would recommend these as excellent texts for reading around the themes of *The Knowledge*:

W. Brian Arthur, *The Nature of Technology: What It Is and How It Evolves*.
George Basalla, *The Evolution of Technology*.
Peter J. Bowler and Iwan Rhys Morus, *Making Modern Science: A Historical Survey*.
Thomas Crump, *A Brief History of Science: As Seen through the Development of Scientific Instruments*.
Patricia Fara, *Science: A Four Thousand Year History*.
John Gribbin, *Science: A History: 1543-2001*.
John Henry, *The Scientific Revolution and the Origins of Modern Science*.
Richard Holmes, *The Age of Wonder: How the Romantic Generation Discovered the Beauty and Terror of Science*.
Steven Johnson, *Where Good Ideas Come From: The Natural History of Innovation*.
Joel Mokyr, *The Lever of Riches: Technological Creativity and Economic Progress*.
Abbott Payson Usher, *A History of Mechanical Inventions*.

Many of the themes of this book, including the conditions of the post-apocalyptic world and recovering from rudimentary means, have also been explored in novels, and here are a few that are well worth recommending. Both Daniel Defoe's *Robinson Crusoe*, Johann David Wyss's *The Swiss Family Robinson*, and Jules Verne's *The Mysterious Island* tell stories of ingenious survival after being knocked back to basics after a shipwreck. Mark Twain's *A Connecticut Yankee in King Arthur's Court* recounts the efforts of an accidental time traveler, and *Island in the Sea of Time* by S. M. Stirling describes how the whole population of an island thrives after being transported back to the Bronze Age by an unexplained event. George R. Stewart's *Earth Abides* follows a community recovering from an apocalypse delivered by plague, whereas John Christopher's *The Death of Grass* covers the catastrophe wrought by disease that doesn't affect humanity directly, but kills all grass species. Cormac McCarthy's *The Road* is a brutal tale of father and son struggling for their lives in the lawless aftermath of an unspecified cataclysm, and Algis Budrys's *Some Will Not Die* and David Brin's *The Postman* deal with the struggle for power after the collapse of civilization, whereas Richard Matheson's *I Am Legend* tells the story of the last surviving human. Pat Frank's *Alas, Babylon* and Nevil Shute's *On the Beach* both describe the immediate aftermath of a nuclear war, whereas *A Canticle for Leibowitz* by Walter M. Miller Jr. considers the preservation of ancient knowledge centuries after a nuclear holocaust. *Riddley Walker* by Russell Hoban also looks at society generations after an apocalypse, but one that has regressed to a nomadic existence. Margaret Atwood's two post-apocalyptic novels, *Oryx and Crake* and *The Year of the Flood*, as well as Jack McDevitt's *Eternity Road* and Kim Stanley Robinson's *The Wild Shore*, also present fascinating visions of life in a post-apocalyptic world. Also well worth reading are the anthologies of post-apocalyptic fiction: *The Ruins of Earth* (edited by Thomas M. Disch), *Wastelands: Stories of the Apocalypse* (ed. John Joseph Adams), and *The Mammoth Book of Apocalyptic SF* (ed. Mike Ashley).

There is also a substantial literature on the alluring beauty of ruins and decaying urban spaces, the topic of the first chapter. Three good recent examples are Andrew Moore's photographs in *Detroit Disassembled*; *Forbidden Places* by Sylvain Margaine; and RomanyWG's *Beauty in Decay*.

I have also provided below a list of a few of the most relevant sources for the general subject matter of each chapter of this book, as well as the references for specific points. Many of these books belong to the Appropriate Technology Library, as indicated by the ATL reference number in brackets after the title. The ATL consists of more than a thousand digitized volumes that have been selected for the practical information they provide on self-sufficiency and rudimentary techniques, and is available on DVD or CD-ROM from Village Earth at http://villageearth.org/appropriate-technology/. Full citation information is given in the bibliography, and also see The Knowledge website, the-knowledge.org, for links to all the cited literature, including free downloads where available.

INTRODUCTION

Nick Bostrom and Milan M. Ćirković, eds., *Global Catastrophic Risks.*
Jared Diamond, *Collapse: How Societies Choose to Fail or Succeed.*
Paul R. Ehrlich and Anne H. Ehrlich, "Can a Collapse of Global Civilization Be Avoided?"
John Michael Greer, *The Long Descent: A User's Guide to the End of the Industrial Age.*
Bob Holmes, "Starting Over: Rebuilding Civilisation from Scratch."
Debora MacKenzie, "Why the Demise of Civilisation May Be Inevitable."
Jeffrey C. Nekola et al., "The Malthusian-Darwinian Dynamic and the Trajectory of Civilization."
Glenn M. Schwartz and John J. Nichols, eds., *After Collapse: The Regeneration of Complex Societies.*
Joseph A. Tainter, *The Collapse of Complex Societies.*

3n **technological regression in Moldova:** Connolly (2001).
4 *I, Pencil:* Read (1958). Also see Ashton (2013), "What Coke Contains."
4 **the Toaster Project:** Thwaites (2011).
7 **a book for all seasons:** Lovelock (1998). Also see the rebuttal to Lovelock's proposal, "How Not to Save Science," in Greer (2006),

as well as more recent proposals for collating and preserving crucial knowledge in Kelly (2006), Raford (2009), Rose (2010), and Kelly (2011), and the humorous essential T-shirt for time travelers at http://www.topatoco.com/bestshirtever.

8 **the encyclopedia as a safe repository of human knowledge:** Yeo (2001).

8 **Apollo program:** http://www.nasa.gov/centers/langley/news/factsheets /Apollo.html.

9 **100 million man-hours devoted to Wikipedia:** Shirky (2010).

9 **Richard Feynman quote:** "Atoms in Motion," Chapter 1 in *The Feynman Lectures on Physics* (1964), now available free at http:// feynmanlectures.caltech.edu/I_01.html.

10 **"These fragments I have shored against my ruins":** T. S. Eliot (1922), *The Waste Land.*

11n **fantasy of starting from scratch:** In addition to the novels *Robinson Crusoe, The Swiss Family Robinson,* and *The Mysterious Island,* already mentioned, a number of other fiction books explore the theme of using crucial knowledge to start over. These include Mark Twain's 1889 novel of an accidental time traveler, *A Connecticut Yankee in King Arthur's Court;* H. G. Wells's 1895 novel *The Time Machine;* and S. M. Stirling's *Island in the Sea of Time* (1998), about an entire modern community transported back to the Bronze Age.

12 **wheelbarrow:** Lewis (1994).

13 **leapfrogging:** Davison (2000), Economist (2006), Economist (2008a, b), McDermott (2010).

13 **Japan leapfrogging:** Mason (1997).

14 **intermediate or appropriate technology:** Rybczynski (1980), Carr (1985).

16 **repurposing:** Edgerton (2007b).

1: THE END OF THE WORLD AS WE KNOW IT

Bruce D. Clayton, *Life After Doomsday: A Survivalist Guide to Nuclear War and Other Major Disasters.*

Aton Edwards, *Preparedness Now! An Emergency Survival Guide.*

Dan Martin, *Apocalypse: How to Survive a Global Crisis.*

James Wesley Rawles, *How to Survive the End of the World as We Know It: Tactics, Techniques and Technologies for Uncertain Times.*

Laura Spinney, "Return to Paradise: If the People Flee, What Will Happen to the Seemingly Indestructible?"

Matthew R. Stein, *When Technology Fails: A Manual for Self-Reliance, Sustainability, and Surviving the Long Emergency.*
Neil Strauss, *Emergency: This Book Will Save Your Life.*
United States Army, *Survival (Field Manual 3-05.70).*
Alan Weisman, *The World Without Us.*
John "Lofty" Wiseman, *SAS Survival Handbook: The Ultimate Guide to Surviving Anywhere.*
Jan Zalasiewicz, *The Earth After Us: What Legacy Will Humans Leave in the Rocks?*

(I should, however, urge caution as not all that is contained in some of the post-appocalyptic survival guides listed above is good advice, particularly in the medical sections.)

19 **Epigraph:** Denis Diderot, *Encyclopédie, ou dictionnaire raisonné des sciences, des arts et des métiers,* 1751–1772. Text available free in the original French from the ARTFL Encyclopédie Project, http:// artflsrv02.uchicago.edu/cgi-bin/philologic/getobject.pl?c.4:1252 .encyclopedie0513. This English translation taken from Yeo (2001).
23n **the Black Death and its social ramifications:** Sherman (2006), Martin (2007).
24 **the theoretical minimum needed for repopulation:** Murray-McIntosh (1998), Hey (2005).
24 **"I Am Legend" scenario:** Richard Matheson, *I Am Legend,* 1954.
26 **recolonization by nature and decay of the cities:** Spinney (1996), Weisman (2007), Zalasiewicz (2008).
31 **the post-apocalyptic climate:** N. Stern (2006), Van Vuuren (2008), Solomon (2009), Cowie (2013).

2: THE GRACE PERIOD
Godfrey Boyle and Peter Harper, *Radical Technology.*
Jim Leckie et al., *More Other Homes and Garbage: Designs for Self-sufficient Living.*
Alexis Madrigal, *Powering the Dream: The History and Promise of Green Technology.*
Nick Rosen, *How to Live Off-Grid: Journeys Outside the System.*
John Seymour, *The New Complete Book of Self-sufficiency.*
Dick and James Strawbridge, *Practical Self Sufficiency: The Complete Guide to Sustainable Living.*
Jon Vogler, *Work from Waste: Recycling Waste to Create Employment.*

33 **Epigraph:** Daniel Defoe, *The Life and Adventures of Robinson Crusoe*, 1719. Text available free from Project Gutenberg, http://www .gutenberg.org/ebooks/521.

33 **prepping and survival during a major crisis:** Clayton (1980), Edwards (2009), Martin (2011), Rawles (2009), Stein (2008), Strauss (2009), United States Army (2002).

36 **water purification:** Huisman (1974), Volunteers in Technical Assistance (1977), Conant (2005).

40 **UK national food reserve:** Department for Environment, Food and Rural Affairs (2010), Department for Environment, Food and Rural Affairs (2012).

42 **degrading GPS accuracy:** pers. comm., U. S. Coast Guard Navigation Center.

43 **how long a stash of medications would last before they expire:** Cohen (2000), Pomerantz (2004).

46 **off-grid electricity:** Leckie (1981), Rosen (2007), Madrigal (2011), Clews (1973).

48 **Goražde jury-rigged hydropower:** Sacco (2000).

49 **rudimentary plastic recycling:** Vogler (1984).

3: AGRICULTURE

Mauro Ambrosoli, *The Wild and the Sown: Botany and Agriculture in Western Europe, 1350–1850.*
Percy Blandford, *Old Farm Tools and Machinery: An Illustrated History.*
Felipe Fernández-Armesto, *Food: A History.*
John Seymour, *The New Complete Book of Self-sufficiency.*
Tom Standage, *An Edible History of Humanity.*

53 **Epigraph:** Wyndham (1951)

58 **soil composition:** P. Stern (1979), Wood (1981).

60 **farm tools:** Blandford (1976), Food and Agriculture Organization of the United Nations (1976), Hurt (1982).

60 **harnessing oxen to a plow:** Starkey (1989).

66 **cereals:** Food and Agriculture Organization of the United Nations (1977).

67 **"humanity subsists, either directly or indirectly, by eating grass":** The potential consequences of this are explored brilliantly in John Christopher's 1956 novel *The Death of Grass*, in which the agent of

doomsday is not a virus that infects humanity, but a plant pathogen wiping out grass species.

73 **Composting:** Gotaas (1976), Dalzell (1981), Shuval (1981), De Decker (2010a).
75 **biogas:** House (1978), Goodall (2008), Strawbridge (2010).
75 **honey-sucker trucks in Bangalore:** Pearce (2013).
75 **Dillo Dirt:** http://austintexas.gov/dillodirt.
76 **superphosphate fertilizer factories in London:** Weisman (2007).
76 **Canadian potash:** Mokyr (1990).
76 **food production trap:** Standage (2010).

4: FOOD AND CLOTHING

Agromisa Foundation, *Preservation of Foods.*
Felipe Fernández-Armesto, *Food: A History.*
Joan Koster, *Handloom Construction: A Practical Guide for the Non-Expert.*
Michael Pollan, *Cooked: A Natural History of Transformation.*
John Seymour, *The New Complete Book of Self-sufficiency.*
Tom Standage, *An Edible History Humanity.*
Carol Hupping Stoner, *Stocking Up: How to Preserve the Foods You Grow Naturally.*
Abbott Payson Usher, *A History of Mechanical Inventions.*

79 **Epigraph:** Quote from "The Ruin," an eighth-century poem in the Exeter Book, written by an unknown Saxon author lamenting Roman ruins. This translation taken from Tainter (1988).
81 **food preservation:** Agromisa Foundation (1990), Buttriss (1999), Stoner (1973).
83 **jury-rigged smokehouse:** Stoner (1973).
84n **nixtamalization:** Fernández-Armesto (2001).
86 **preparation of cereals:** UNIFEM (1988).
88 **preparing a sourdough:** Avery (2001a, b), Lang (2003).
91 **Mongolian still:** Sella (2012).
93 **Zeer pot:** Löfström (2011).
94 **Einstein's refrigerator:** Silverman (2001), Jha (2008).
94 **compressor and absorber designs of refrigerator:** Cowan (1985), Bell (2011).
96 **wool spinning:** Wigginton (1973).
98 **simple weaving:** Koster (1979).

100   the button: Mokyr (1990), Mortimer (2008).
101   mechanization of spinning and weaving: Usher (1982), Mokyr (1990), Allen (2009).

5: SUBSTANCES

Alan P. Dalton, *Chemicals from Biological Resources*.
William B. Dick, *Dick's Encyclopedia of Practical Receipts and Processes*.
Kevin M. Dunn, *Caveman Chemistry: 28 Projects, from the Creation of Fire to the Production of Plastics*.

103   Epigraph: Atwood (2003).
104   thermal energy through history: De Decker (2011).
105   importance of coke in Industrial Revolution: Allen (2009).
106   coppicing firewood: Stanford (1976).
107   charcoal: Goodall (2008).
107n Brazilian charcoal for steel production: Kato et al. (2005).
108n reserve technologies: Edgerton (2007a).
110   lime burning: Wingate (1985).
112   hand washing and reduction of gastrointestinal diseases: Bloomfield and Nath (2009).
114   the importance of alkalis throughout history: Deighton (1907), Reilly (1951).
116   wood pyrolysis: Dumesny and Noyer (1908), Dalton (1973), Boyle and Harper (1976), McClure (2000).
118   acetone shortage in the First World War: David (2012).
120   sulfuric acid: McKee and Salk (1924), Karpenko (2002).

6: MATERIALS

Kevin M. Dunn, *Caveman Chemistry: 28 Projects, from the Creation of Fire to the Production of Plastics*.
Albert Jackson and David Day, *Tools and How to Use Them: An Illustrated Encyclopedia*.
Carl G. Johnson and William R. Weeks, *Metallurgy*.
Richard Shelton Kirby et al., *Engineering in History*.

123   Epigraph: Miller (1960).
123   wood: United States Department of Agriculture (1974).
124   basic construction techniques: Leckie (1981), P. Stern (1983), van Lengen (2008).

127 Roman pozzolana cement: Oleson (2008).
129 reinforced concrete: P. Stern (1983).
130 ironworking: Weygers (1974), Winden (1990).
130 making and using tools: Weygers (1973), Jackson and Day (1978).
131 hardening and tempering tools: Gentry and Westbury (1980).
131 oxyacetylene torch: Parkin and Flood (1969).
131 arc welding: Lincoln Electric Company (1973).
132 small-scale foundry and metal casting: Aspin (1975).
132 bootstrapping a complete metalworking shop: Gingery (2011).
132 iron smelting: Allen (2009), Johnson (1977).
136 Chinese blast furnace: Mokyr (1990).
137 Bessemer process: Mokyr (1990).
138 glassmaking: Whitby (1983).
140 "lead crystal" glass: MacLeod (1987).
141 central role of glass in science: Macfarlane (2002).

7: MEDICINE
Murray Dickson, *Where There Is No Dentist*.
Roy Porter, *Blood and Guts: A Short History of Medicine*.
Anne Rooney, *The Story of Medicine: From Early Healing to the Miracles of Modern Medicine*.
David Werner, *Where There Is No Doctor: A Village Health Care Handbook*.

145 Epigraph: John Lloyd Stephens, ca. 1841, in Diamond, *Collapse* (2005).
146 diseases from animals: Porter (2002), Rooney (2009).
147 importance of sanitation: Solomon (2011), Conant (2005), Mann and Williamson (1982).
148 cholera: Clark (2010).
148 oral rehydration therapy: Conant (2005).
149n obstetric forceps kept a secret: Porter (2002).
150 car-parts incubator: Johnson (2010), http://designthatmatters.org/impact/ heonurture.
153 serendipitous discovery of X-rays: Gribbin (2002), Osman (2011), Kean (2010).
157 willow bark and aspirin: Mokyr (1990), Pollard (2010).
157n scurvy and the first clinical trial: Osman (2011).
158 principles of surgery: Cook (1988).
158 anesthetics: Dobson (1988).
159 nitrous oxide: Gribbin (2002), Holmes (2008).

160 instructions for rudimentary microscope: Casselman (2011).

160n Leeuwenhoek: Macfarlane (2002), Crump (2001), Gribbin, (2002), Sherman (2006).

161n Marcus Terentius Varro: Rooney (2009).

162 serendipitous discovery of antibiotics: Lax (2004), Kelly (2010), Winston (2010), Pollard (2010).

163 extraction and mass production of penicillin: Lax (2004).

8: POWER TO THE PEOPLE

Godfrey Boyle and Peter Harper, *Radical Technology*.

Alexis Madrigal, *Powering the Dream: The History and Promise of Green Technology*.

Abbott Payson Usher, *A History of Mechanical Inventions*.

165 Epigraph: Frank (1959).

167 Roman waterwheel: Usher (1982), Oleson (2008).

168 key innovations during the supposedly "dark" medieval era: Fara (2009).

168 windmills: McGuigan (1978a), Mokyr (1990), Hills (1996), De Decker (2009.)

171 importance of waterwheels and windmills: Basalla (1988).

171 diverse uses of waterwheels and windmills: Usher (1982), Solomon (2011).

171 mechanisms for transforming motion: Hiscox (2007), Brown (2008).

172 suction pumps: Fraenkel (1997).

172 steam engine: Usher (1982), Mokyr (1990), Crump (2001), Allen (2009).

175 voltaic pile: Gribbin (2002).

175 Baghdad battery: Schlesinger (2010), Osman (2011).

176 discovery of electromagnetism: Crump (2001), Gribbin (2002), Hamilton (2003), Fara (2009), Schlesinger (2010), Ball (2012).

179 retrofitting a traditional four-sail windmill: Watson and Thomson (2005).

179 Charles Brush's electricity-generating windmill: Hills (1996), Winston (2010), Krouse (2011).

180 water turbines: McGuigan (1978b), Usher (1982), Holland (1986), Mokyr (1990), Eisenring (1991).

9: TRANSPORT

187 Epigraph: Dahl (1975).

189 Rudolf Diesel quote: Goodall (2008).

190 **bioethanol:** United States Department of Energy (1980), Goodall (2008).

190 **biodiesel:** Rosen (2007), Strawbridge (2010).

191 **gas bag vehicles:** House (1978), De Decker (2011b).

193 **wood-fueled Tiger tanks:** Krammer (1978).

193 **wood gasifiers:** De Decker (2010b), Food and Agriculture Organization of the United Nations (1986), LaFontaine and Zimmerman (1989).

194 **guayule:** National Academy of Sciences (1977).

195 **harnessing oxen:** Starkey (1989).

196 **throat-and-girth harness and horse collar:** Mokyr (1990).

197 **peak horse use:** Edgerton (2007a, b).

198 **sails:** Farndon (2010).

198n **Cuban resurrection of animal traction:** Edgerton (2007b).

200 **penny-farthing and modern safety bicycle:** Broers (2005).

201 **the nature of novel technologies and the automobile as a lashing together of preexisting mechanical solutions:** Arthur (2009), Kelly (2010), Mokyr (1990).

201 **internal combustion engine and motor vehicle mechanisms:** Bureau of Naval Personnel (1971), Hillier and Pittuck (1981), Usher (1982).

207 **history of electric cars:** Crump (2001), Edgerton (2007a), Brooks (2009), De Decker (2010c), Madrigal (2011).

10: COMMUNICATION

J. P. Davidson, *Planet Word: The Story of Language from the Earliest Grunts to Twitter and Beyond.*

209 **Epigraph:** Shelley (1818).

210 **history of paper:** Mokyr (1990).

211 **chemical liberation of cellulose fibers:** Dunn (2003).

212 **papermaking:** Vigneault (2007), Seymour (2009).

213 **ink from berries:** Dragotta (2007).

214 **iron gall ink:** Finlay (2002), Fruen (2002), Smith (2009).

215 **social ramifications of printing press:** Broers (2005), Farndon (2010).

215 **development of printing press:** Usher (1982), Mokyr (1990), Finlay (2002), Johnson (2010).

222 **rudimentary radio transmitters and receivers:** Parker (1999), Field (n. d.), Crump (2001).

226 foxhole/POW radios: Wells (1995), Ross (2005), Carusella (2008), and see Gillies (2011) for further ingenuity among POWs.

11: ADVANCED CHEMISTRY

Kevin M. Dunn, *Caveman Chemistry: 28 Projects, from the Creation of Fire to the Production of Plastics.*

Sam Kean, *The Disappearing Spoon: And Other True Tales of Madness, Love, and the History of the World from the Periodic Table of the Elements.*

Joel Mokyr, *The Lever of Riches: Technological Creativity and Economic Progress.*

231 **Epigraph:** Coupland (1992).
232 **electrolysis of seawater:** Abdel-Aal et al. (2010).
232 **aluminum:** Johnson (1977), Kean (2010).
233 **electrolysis and discovery of new elements:** Gribbin (2002), Holmes (2008).
234 **periodic table:** Fara (2009), Kean (2010).
236 **black powder as elixir for immortality:** Winston (2010).
238 **nitroglycerin and dynamite:** Mokyr (1990).
239 **applications of photography:** Gribbin (2002), Osman (2011).
240 **rudimentary photography:** Sutton (1986), Ware (1997), Crump (2001), Ware (2002), Ware (2004).
242 **industrial chemistry:** Mokyr (1990).
243 **demand for soda:** Deighton (1907), Reilly (1951).
244 **Leblanc process, early industrial pollution, Solvay processes:** Deighton (1907), Reilly (1951), Mokyr (1990).
248 **William Crookes quote:** Standage (2010).
248 **nitrogen gas is the least reactive diatomic substance:** Schrock (2006).
248 **Haber-Bosch process:** Standage (2010), Kean (2010), Perkins (1977), Edgerton (2007a).

12: TIME AND PLACE

Adam Frank, *About Time: Cosmology and Culture at the Twilight of the Big Bang.*

Eric Bruton, *The History of Clocks & Watches.*

Dava Sobel, *Longitude: The True Story of a Lone Genius Who Solved the Greatest Scientific Problem of His Time.*

253 Epigraph: Denis Diderot as quoted by Goodman (1995).
254 constancy of sand time (hourglass) compared to water clock: Bruton (2000).
256 sundials: Oleson (2008).
256n Manhattan as a city-size Stonehenge: Astronomy Picture of the Day, July 12, 2006, http://apod.nasa.gov/apod/apo60712.html.
257 mechanical clocks: Usher (1982), Bruton (2000), Gribbin (2002), Frank (2011).
258 60 seconds, 60 minutes, 24 hours: Crump (2001), Frank (2011).
259 "o'clock": Mortimer (2008).
261 first appearance of Sirius: Schaefer (2000).
262 resurrect the Gregorian calendar: see Pappas (2011) for one proposal for reformatting the year into a different structure of months.
270 navigation before accurate clocks by sailing along line of latitude: Usher (1982).
271 solving the longitude problem: Sobel (1995).
271 spring-based clocks: Usher (1982), Bruton (2000).
272n 22 chronometers aboard HMS *Beagle*: Sobel (1995).

13: THE GREATEST INVENTION
275 Epigraph: Eliot (1943).
276 nothing inevitable about technological progress and history of China: Mokyr (1990).
277 Industrial Revolution in eighteenth-century Britain: Allen (2009).
282n metric system and why UK and USA did not adopt it: Crump (2001).
284 invention of barometer and thermometer: Crump (2001), Chang (2004).
286 the scientific revolution and how science is done: Shapin (1996), Kuhn (1996), Bowler and Morus (2005), Henry (2008), Ball (2012).
289 symbiosis between science and technology: Basalla (1988), Mokyr (1990), Bowler and Morus (2005), Arthur (2009), Johnson (2010).

Abdel-Aal, H.K., K.M. Zohdy, and M. Abdel Kareem. 2010. "Hydrogen Production Using Sea Water Electrolysis." *The Open Fuel Cells Journal* 3: 1-7.

Adams, John Joseph, ed. 2008. *Wastelands: Stories of the Apocalypse*. San Francisco: Night Shade Books.

Agromisa Human Nutrition and Food Processing Group. 1990. *Preservation of Foods*. (ATL 07-289) Wageningen, Netherlands: Agromisa Foundation.

Ahuja, Rajeev, Andreas Blomqvist, Peter Larsson, et al. 2011. "Relativity and the Lead-acid Battery." *Physical Review Letters* 106 (1): 018301.

Allen, Robert C. 2009. *The British Industrial Revolution in Global Perspective*. Cambridge, UK: Cambridge University Press.

Ambrosoli, Mauro. 1997. *The Wild and the Sown: Botany and Agriculture in Western Europe, 1350-1850*. Past and Present Publications. Cambridge, UK: Cambridge University Press. Originally published as *Scienziati, contadini e proprietari: botanica e agricoltura nell'Europa oeridentale 1350–1850* (Turin: Giulio Einaudi, 1992).

Arthur, W. Brian. 2009. *The Nature of Technology: What It Is and How It Evolves*. New York: Free Press.

Ashton, Kevin. 2013. "What Coke Contains." *Medium*. April 5. https://medium.com/the-ingredients-2/221d449929ef.

Aspin, B. Terry. 1975. *Foundrywork for the Amateur*. (ATL 04-94) Hemel Hempstead, UK: Model and Ailied Publications, Argus Books.

Atwood, Margaret. 2003. *Oryx and Crake*. Toronto: McClelland and Stewart.

Avery, Mike. 2001a. "What Is Sourdough?" SourdoughHome, http://www.sourdoughhome.com/index.php?content=whatissourdough.

———. 2001b. "Starting a Starter." SourdoughHome, http://www.sourdoughhome.com/index.php?content=startermyway2.

Ball, Philip. 2012. *Curiosity: How Science Became Interested in Everything*. Chicago and London: The University of Chicago Press.

Basalla, George. 1988. *The Evolution of Technology*. Cambridge History of Science Series. Cambridge UK: Cambridge University Press.

Bell, Alice. 2011. "How the Refrigerator Got Its Hum." *through the looking glass*

(blog), September 19. http://alicerosebell.wordpress.com/2011/09/19/how-the
-refrigerator-got-its-hum/.

Blandford, Percy. 1976. *Old Farm Tools and Machinery: An Illustrated History*.
Newton Abbot, UK: David & Charles.

Bloomfield, Sally F., and Kumar Jyoti Nath. 2009. *Use of Ash and Mud for
Handwashing in Low Income Communities*. Mantacute, UK: International
Scientific Forum on Home Hygiene.

Bostrom, Nick, and Milan M. Ćirković, eds. 2008. *Global Catastrophic Risks*.
Oxford and New York: Oxford University Press.

Bowler, Peter J., and Iwan Rhys Morus. 2005. *Making Modern Science: A Historical
Survey*. Chicago: The University of Chicago Press.

Boyle, Godfrey, Peter Harper, and the editors of *Undercurrents*. 1976. *Radical
Technology: Food, Shelter, Tools, Materials, Energy, Communication, Autonomy,
Community*. (ATL 01-13) New York: Pantheon Books.

Buttriss, Judith. 1999. *Nutrition and Food Processing*. London: British Nutrition
Foundation.

Broers, Alec. 2005. *The Triumph of Technology: The BBC Reith Lectures 2005*.
Cambridge UK: Cambridge University Press.

Brooks, Michael. 2009. "Electric Cars: Juiced Up and Ready to Go." *New Scientist*,
July 20: 42–45.

Brown, Henry T. 2008. *507 Mechanical Movements: Mechanisms and Devices*. 18th ed.
(1906). La Vergne, TN: BN Publishing. First published 1868.

Bruton, Eric. 2000. *The History of Clocks & Watches*. London: Little, Brown.

Bureau of Naval Personnel. 1971. *Basic Machines and How They Work*. (ATL 04-81)
Mineola, NY: Dover Publications.

Carr, Marilyn, ed. 1985. *The AT Reader: Theory and Practice in Appropriate
Technology*. (ATL 01-20) ITDG Publishing.

Carusella, Brian. 2008. "Foxhole and PoW Built Radios: History and
Construction." Bizarre Stuff You Can Make in Your Kitchen. http://bizarrelabs
.com/foxhole.htm.

Casselman, Anne. 2011. "Microscope, DIY, 3 Minutes." *The Last Word on Nothing*
(blog), September 5. http://www.lastwordonnothing.com/2011/09/05/guest-post
-microscope-diy/.

Chang, Hasok. 2004. *Inventing Temperature: Measurement and Scientific Progress*.
Oxford and New York: Oxford University Press.

Clark, David P. 2010. *Germs, Genes & Civilization: How Epidemics Shaped Who We
Are Today*. Upper Saddle River, NJ: FT Press / Pearson.

Clayton, Bruce D. 1980. *Life After Doomsday: A Survivalist Guide to Nuclear War and
Other Major Disasters*. Boulder, CO: Paladin Press.

Clews, Henry. 1974. *Electric Power from the Wind*. (ATL 21-466) Norwich, VT:
Enertech Corporation.

Cohen, Laurie P. 2000. "Many Medicines Are Potent Years Past Expiration Dates." *Wall Street Journal*, March 28. http://online.wsj.com/article/SB95420150853006 7326.html.

Collins, H. M. 1974. "The TEA Set: Tacit Knowledge and Scientific Networks." *Science Studies* 4 (2): 165–86.

Conant, Jeff. 2005. *Sanitation and Cleanliness for a Healthy Environment*. Berkeley, CA: The Hesperian Foundation.

Connolly, Kate. 2001. "Human Flesh On Sale in Land the Cold War Left Behind." *The Observer*, April 7. http://www.theguardian.com/world/2001/apr/08/russia .kateconnolly.

Cook, John, Balu Sankaran, and Ambrose E.O. Wasunna (eds.). 1988. *General Surgery at the District Hospital*. (ATL 27-721) Geneva: World Health Organization.

Coupland, Douglas. 1992. *Shampoo Planet*. New York: Pocket Books.

———. 1998. *Girlfriend in a Coma*. New York: Regan Books / HarperCollins.

Cowan, Ruth Schwartz. 1985. "How the Refrigerator Got Its Hum." In *The Social Shaping of Technology*, edited by Donald MacKenzie and Judy Wajcman. Milton Keynes, UK, and Philadelphia: Open University Press.

Cowie, Jonathan. 2012. *Climate Change: Biological and Human Aspects*. 2nd ed. Cambridge, UK: Cambridge University Press.

Crump, Thomas. 2001. *A Brief History of Science: As Seen through the Development of Scientific Instruments*. London: Constable & Robinson.

Dahl, Roald. 1975. *Danny, the Champion of the World*. London: Jonathan Cape.

Dalton, Alan P. 1973. *Chemicals from Biological Resources*. London: Intermediate Technology Development Group.

Dalzell, Howard W., Kenneth R. Gray, and A.J. Biddlestone. 1981. *Composting in Tropical Agriculture*. (ATL 05-165) Ashford, UK: International Institute of Biological Husbandry.

David, Saul. 2012. "How Germany Lost the WWI Arms Race." BBC News Magazine, February 15. http://www.bbc.co.uk/news/magazine-17011607.

Davidson, J. P. 2011. *Planet Word: The Story of Language from the Earliest Grunts to Twitter and Beyond*. To accompany the BBC series. London: Michael Joseph / Penguin.

Davison, Robert, Doug Vogel, Roger Harris, and Noel Jones. 2000. "Technology Leapfrogging in Developing Countries: An Inevitable Luxury?" *The Electronic Journal of Information Systems in Developing Countries* 1 (5): 1–10.

De Decker, Kris. 2009. "Wind Powered Factories: History (and Future) of Industrial Windmills." *Low-tech Magazine*, October 8. http://www .lowtechmagazine.com/2009/10/history-of-industrial-windmills.html.

———. 2010a. "Recycling Animal and Human Dung Is the Key to Sustainable Farming." *Low-tech Magazine*, September 15. http://www.lowtechmagazine

.com/2010/09/recycling-animal-and-human-dung-is-the-key-to-sustainable
-farming.html.

⸻. 2010b. "Wood Gas Vehicles: Firewood in the Fuel Tank." *Low-tech
Magazine*, January 18. http://www.lowtechmagazine.com/2010/01/wood-gas
-cars.html.

⸻. 2010c. "The Status Quo of Electric Cars: Better Batteries, Same Range."
Low-tech Magazine, May 3. http://www.lowtechmagazine.com/2010/05/the
-status-quo-of-electric-cars-better-batteries-same-range.html.

⸻. 2011a. "Medieval Smokestacks: Fossil Fuels in Pre-industrial Times."
Low-tech Magazine, September 29. http://www.lowtechmagazine.com/2011
/09/peat-and-coal-fossil-fuels-in-pre-industrial-times.html.

⸻. 2011b. "Gas Bag Vehicles." *Low-tech Magazine*, November 13. http://www
.lowtechmagazine.com/2011/11/gas-bag-vehicles.html.

Deighton, T. Howard. 1907. *The Struggle for Supremacy: Being a Series of Chapters in
the History of the Leblanc Alkali Industry in Great Britain*. Liverpool: Gilbert G.
Walmsley.

Department for Environment, Food and Rural Affairs. 2010. *UK Food Security
Assessment: Detailed Analysis*. London: DEFRA.

⸻. 2012. *Food Statistics Pocketbook*. London: Department for Environment,
Food and Rural Affairs.

Diamond, Jared. 2005. *Collapse: How Societies Choose to Fail or Survive*. New York:
Viking Penguin.

Dick, William B. 1872. *Dick's Encyclopedia of Practical Receipts and Processes*. (ATL
02-26) New York: Dick & Fitzgerald. Reprint, New York: Funk & Wagnalls,
1975.

Dickson, Murray. 2011. *Where There Is No Dentist*. Berkeley, CA: The Foundation
Hesperian.

Dobson, Michael B. 1988. *Anesthesia at the District Hospital*. (ATL 27-720) Geneva:
World Health Organization.

Dragotta, Nick. 2007. "Pen Pal." *Craft* (Winter). www.howtoons./?page_id-134.

Dumesny, P. and J. Noyer. 1908. *Wood Products: Distillates and Extracts*. Translated
by Donald Grant. New York: D. van Nostrand Co. Reprint 2007, Knowledge
Publications.

Dunn, Kevin M. 2003. *Caveman Chemistry: 28 Projects, from the Creation of Fire to
the Production of Plastics*. Boca Raton, FL: Universal Publishers.

Economist, The. 2006. "Behind the Bleeding Edge: Skipping over old technologies
to adopt new ones offers opportunities—and a lesson." September 21. http://
www.economist.com/node/7944359.

⸻. 2008a. "Of Internet Cafés and Power Cuts: Emerging economies are better
at adopting new technologies than at putting them into widespread use."
February 7. http://www.economist.com/node/10640716.

————. 2008b. "The Limits of Leapfrogging: The spread of new technologies often depends on the availability of older ones." February 7. http://www.economist .com/node/10650775.

————. 2012. "Doomsdays: Predicting the End of the World." December 20. economist.com/blogs/graphicdetail/2012/12/daily-chart-11.

Edgerton, David. 2007a. *The Shock of the Old: Technology and Global History Since 1900.* New York: Oxford University Press.

————. 2007b. "Creole Technologies and Global Histories: Rethinking How Things Travel in Space and Time." *Journal of History of Science and Technology* 1 (Summer): 75–112.

Edwards, Aton. 2009. *Preparedness Now! An Emergency Survival Guide.* Exp. and rev. ed. Process Self-Reliance. Port Townsend, WA: Process Media.

Ehrlich, Paul R., and Anne H. Ehrlich. 2013. "Can a Collapse of Global Civilization be Avoided?" *Proceedings of the Royal Society B* 280 (1754): 20122845.

Eisenring, Markus. 1991. *Micro Pelton Turbines.* (ATL 22-543) St. Gallen, Switzerland: SKAT, Swiss Center for Appropriate Technology.

Eliot, T. S. 1943. *Four Quartets.* New York: Harcourt, Brace & Company.

Food and Agriculture Organization of the United Nations. 1976. *Farming with Animal Power.* (ATL 05-150) Better Farming Series 14. Rome: FAO.

Food and Agriculture Organization of the United Nations. 1977. *Cereals.* (ATL 05-151) Better Farming Series 15. Rome: FAO.

Food and Agriculture Organization of the United Nations, Forestry Department. 1986. *Wood Gas as Engine Fuel.* Rome: FAO.

Fara, Patricia. 2009. *Science: A Four Thousand Year History.* Oxford and New York: Oxford University Press.

Farndon, John. 2010. *The World's Greatest Idea: The Fifty Greatest Ideas that Have Changed Humanity.* London: Icon Books.

Ferguson, Niall. 2011. *Civilization: The West and the Rest.* London: Allen Lane / Penguin.

Fernández-Armesto, Felipe. 2001. *Food: A History.* London: Macmillan. (Published in the US as *Near a Thousand Tables: A History of Food.* New York: The Free Press, 2002.)

Field, Simon Quellen. n. d. "Building a Crystal Radio Out of Household Items." Scitoys: Science Toys You Can Make with Your Kids. http://scitoys.com/scitoys/ scitoys/radio/homemade_radio.html.

Finlay, Victoria. 2002. *Colour: Travels Through the Paintbox.* London: Hodder and Stoughton. US edition: *Color: A Natural History of the Palette.* New York: Random House, 2003.

Frank, Adam. 2011. *About Time: Cosmology and Culture at the Twilight of the Big Bang.* New York: Free Press.

Frank, Pat. 1959. *Alas, Babylon.* New York: J. B. Lippincott.

Frankael, Peter. 1997. *Water-Pumping Devices: A Handbook for Users and Choosers* (ATL 14-370). London: Intermediate Technology Publications.

Fruen, Lois. 2002. "Iron Gall Ink." A teaching resource to accompany the author's textbook *The Real World of Chemistry*, 6th ed. (Dubuque, IA: Kendall Hunt Publishing). http://www.realscience.breckschool.org/upper/fruen/files /Enrichmentarticles/files/IronGallInk/IronGallInk.html.

Gentry, George, and Edgar T. Westbury. 1980. *Hardening and Tempering Engineers' Tools*. 3rd ed. (ATL 04-98) Watford, UK: Model and Allied Publications, Argus Books.

Gillies, Midge. 2011. *The Barbed-wire University: The Real Lives of Prisoners of War in the Second World War*. London: Aurum Press.

Gingery, David J., and Vincent R. Gingery. 2011. *Build Your Own Metal Working Shop from Scrap (Complete 7 Book Series)*. Rogersville, MO: David J. Gingery Publishing.

Goodall, Chris. 2008. *Ten Technologies to Fix Energy and Climate*. London: Profile Books.

Goodman, John, ed. 1995. *Diderot on Art: II: The Salon of 1767*. New Haven, CT, and London: Yale University Press.

Gotaas, Harold B. 1956. *Composting: Sanitary Disposal and Reclamation of Organic Wastes*. (ATL 05-166) Geneva: World Health Organization.

Greer, John Michael. 2006. "How Not To Save Science." *The Archdruid Report* (blog). July 26. http://thearchdruidreport.blogspot.co.uk/2006/07/how-not-to -save-science.html.

———. 2008. *The Long Descent: A User's Guide to the End of the Industrial Age*. Gabriola Island, BC, Canada: New Society Publishers.

Gribbin, John. 2002. *Science: A History: 1543-2001*. London: Allen Lane / Penguin. New York: Penguin Books, 2003.

Hamilton, James. 2003. *Faraday: The Life*. London: HarperCollins.

Henry, John. 2008. *The Scientific Revolution and the Origins of Modern Science*. 3rd ed. Studies in European History. New York: Palgrave Macmillan.

Hey, Jody. 2005. "On the Number of New World Founders: A Population Genetic Portrait of the Peopling of the Americas." *PLoS Biology* 3 (6): e193.

Hillier, V. A. W., and F. Pittuck. 1981. *Fundamentals of Motor Vehicle Technology*. 3rd ed. London: Hutchinson Radius.

Hills, Richard L. 1996. *Power from Wind: A History of Windmill Technology*. Cambridge, UK: Cambridge University Press.

Hiscox, Gardner Dexter. 2007. *1800 Mechanical Movements, Devices and Appliances*. Dover Science Books. Mineola, NY: Dover Publications.

Holland, Ray. 1983. *Micro Hydro Electric Power*. (ATL 22-531) London: Intermediate Technology Development Group.

Holmes, Bob. 2011. "Starting Over: Rebuilding Civilisation from Scratch." *New Scientist*, March 28: 40–45.

Holmes, Richard. 2008. *The Age of Wonder: How the Romantic Generation Discovered the Beauty and Terror of Science.* New York: Pantheon Books.

House, David. 1978. *The Biogas Handbook.* (ATL 24-568) Los Angeles: Peace Press. Rev. ed. published 2006, House Press.

Huisman, L., and W. E. Wood. 1974. *Slow Sand Filtration.* (ATL 16-376) Geneva: World Health Organization.

Hurt, R. Douglas. 1982. *American Farm Tools: From Hand-Power to Steam-Power.* (ATL 06-262) Manhattan, KS: Sunflower University Press.

Jackson, Albert, and David Day. 1978. *Tools and How to Use Them: An Illustrated Encyclopedia.* (ATL 04-122) New York: Alfred A. Knopf.

Jha, Alok. 2008. "Einstein Fridge Design Can Help Global Cooling." *The Observer,* September 20. http://www.theguardian.com/science/2008/sep/21/scienceof climatechange.climatechange.

Johnson, Carl G., and William R. Weeks. 1977. *Metallurgy.* 5th ed. (ATL 04-106) Chicago, IL: American Technical Publishers.

Johnson, Steven. 2010. *Where Good Ideas Come From: The Natural History of Innovation.* New York: Riverhead Books.

Karpenko, Vladimír, and John A. Norris. 2002. "Vitriol in the History of Chemistry." *Chemické Listy* 96: 997–1005.

Kato, M., D. M. DeMarini, A. B. Carvalho, et al. 2005. "World at Work: Charcoal Producing Industries in Northeastern Brazil." *Occupational and Environmental Medicine* 62 (2): 128–32.

Kean, Sam. 2010. *The Disappearing Spoon: And Other True Tales of Madness, Love, and the History of the World from the Periodic Table of the Elements.* New York: Little, Brown and Company.

Kelly, Kevin. 2006. "The Forever Book." *The Technium* (blog), February 22. http://www.kk.org/thetechnium/archives/2006/02/the_forever_boo.php.

———. 2010. *What Technology Wants.* New York: Viking.

———. 2011. "The Library of Utility." *Blog of the Long Now,* April 25. http://blog.longnow.org/02011/04/25/the-library-of-utility/.

Kirby, Richard Shelton, Sidney Withington, Arthur Burr Darling, and Frederick Gridley Kilgour. 1990. *Engineering in History.* Dover Civil and Mechanical Engineering. Mineola, NY: Dover Publications.

Koster, Joan. 1979. *Handloom Construction: A Practical Guide for the Non-Expert.* (ATL 33-778) Arlington, VA: Volunteers in Technical Assistance.

Krammer, Arnold. 1978. "Fueling the Third Reich." *Technology and Culture* 19 (3): 394–422.

Krouse, Peter. 2011. "Charles Brush Used Wind Power in House 120 Years Ago: Cleveland Innovations." *Plain Dealer,* August 11. http://blog.cleveland.com/metro/2011/08/charles_brush_used_wind_power.html.

Kuhn, Thomas S. 1996. *The Structure of Scientific Revolutions.* 3rd ed. Chicago: The University of Chicago Press.

LaFontaine, H., and F. P. Zimmerman. 1989. *Construction of a Simplified Wood Gas Generator for Fueling Internal Combustion Engines in a Petroleum Emergency.* Washington, DC: Federal Emergency Management Agency.

Lang, Jack. 2003. "Sourdough Bread." eGullet Forums, September 10. Society for Culinary Arts and Letters. http://forums.egullet.org/topic/27634-sourdough -bread/.

Lax, Eric. 2004. *The Mold in Dr. Florey's Coat: The Remarkable True Story of the Penicillin Miracle.* New York: Henry Holt and Company.

Leckie, Jim, Gil Masters, Harry Whitehouse, and Lily Young. 1981. *More Other Homes and Garbage: Designs for Self-sufficient Living.* (ATL 02-47) San Francisco: Sierra Club Books.

Lewis, M. J. T. 1994. "The Origins of the Wheelbarrow." *Technology and Culture* 35 (3): 453–75.

Lincoln Electric Company. 1973. *The Procedure Handbook of Arc Welding.* (ATL 04-115) Cleveland, Ohio: The Lincoln Electric Company.

Lisboa, Maria Manuel. 2011. *The End of the World: Apocalypse and Its Aftermath in Western Culture.* Cambridge, UK: Open Book Publishers.

Löfström, Johan. 2011. "Zeer Pot Refrigerator." *Appropedia.* http://www.appropedia .org/Zeer_pot_refrigerator.

Lovelock, James. 1998. "A Book for All Seasons." *Science* 280 (5365): 832–33.

Macfarlane, Alan, and Gerry Martin. 2002. *The Glass Bathyscaphe: How Glass Changed the World.* London: Profile Books.

MacGregor, Neil. 2011. *A History of the World in 100 Objects.* New York: Viking.

MacKenzie, Debora. 2008. "Why the Demise of Civilization May Be Inevitable." *New Scientist*, April 2: 32–35

MacLeod, Christine. 1987. "Accident or Design? George Ravenscroft's Patent and the Invention of Lead-Crystal Glass." *Technology and Culture.* 28 (4): 776–803.

Madrigal, Alexis. 2011. *Powering the Dream: The History and Promise of Green Technology.* Cambridge, MA: Da Capo Press.

Mann, Henry Thomas, and David Williamson. 1982. *Water Treatment and Sanitation: Simple Methods for Rural Areas.* Rev. ed. (ATL 16-381) Bourton on Dunsmore, UK: Practical Action Publishing.

Martin, Dan. 2011. *Apocalypse: How to Survive a Global Crisis.* Special 2012 Edition. Sandy, UT: ECKO House Publishing.

Martin, Felix. 2013. *Money: The Unauthorized Biography.* London: The Bodley Head.

Martin, Sean. 2007. *The Black Death.* Edison, NJ: Chartwell Books.

Mason, R. H. P., and J. G. Caiger. 1997. *A History of Japan.* Rev. ed Rutland, VT, and Tokyo: Charles E. Tuttle Company.

McClure, David Courtney. 2000. "Kilkerran Pyroligneous Acid Works 1845 to

1945." Ayrshire History. http://www.ayrshirehistory.org.uk/AcidWorks /acidworks.htm.

McDermott, Mat. 2010. "Techo-Leapfrogging at Its Best: 2,000 Indian Villages Skip Fossil Fuels, Get First Electricity from Solar." *TreeHugger*, September 22. http://www.trechugger.com/natural-sciences/techo-leapfrogging-at-its-best -2000-indian-villages-skip-fossil-fuels-get-first-electricity-from-solar.html.

McGuigan, Dermot. 1978a. *Small Scale Wind Power*. Dorset, UK: Prism Press.

———. 1978b. *Harnessing Water Power for Home Energy*. (ATL 22-507) Charlotte, VT: Garden Way Publishing.

McKee, Ralph H., and Carroll M. Salls. 1924. "Sulfuryl Chloride: Principles of Manufacture from Sulfur Burner Gas." *Industrial and Engineering Chemistry* 16 (4): 351–53.

Miller, Walter M., Jr. 1959. *A Canticle for Leibowitz*. New York: J. B. Lippincott.

Mokyr, Joel. 1990. *The Lever of Riches: Technological Creativity and Economic Progress*. New York: Oxford University Press.

Mortimer, Ian. 2008. *The Time Traveler's Guide to Medieval England: A Handbook for Visitors to the Fourteenth Century*. London: The Bodley Head; New York: Touchstone, 2009.

Murray-McIntosh, Rosalind P., Brian J. Scrimshaw, Peter J. Hatfield, and David Penny. 1998. "Testing Migration Patterns and Estimating Founding Population Size in Polynesia by Using Human mtDNA Sequences." *Proceedings of the National Academy of Sciences* 95 (15): 9047–52.

National Academy of Sciences. 1977. *Guayule: An Alternative Source of Natural Rubber*. (ATL 05-183) Washington, DC: National Academy of Sciences.

Nekola, Jeffrey C., Craig D. Allen, James H. Brown, et al. 2013. "The Malthusian-Darwinian Dynamic and the Trajectory of Civilization." *Trends in Ecology & Evolution* 28 (3): 127–30.

Oleson, John Peter, ed. 2008. *The Oxford Handbook of Engineering and Technology in the Classical World*. Oxford and New York: Oxford University Press.

Osman, Jheni. 2011. *100 Ideas That Changed the World*. London: BBC Books.

Pappas, Stephanie. 2011. "Is It Time to Overhaul the Calendar?" *Scientific American*, December 29. http://www.scientificamerican.com/article .cfm?id=is-it-time-to-overhaul

Parker, Bev. 1999. "Early Transmitters and Receivers." Wolverhampton History & Heritage Website. http://www.historywebsite.co.uk/Museum/Engineering /Electronics/history/earlytxrx.htm

Parkin, N., and C. R. Flood. 1969. *Welding Craft Practice: Part 1, Volume 1: Oxy-acetylene Gas Welding and Related Studies*. (ATL 04-126) Pergamon International Library of Science, Technology, Engineering and Social Studies. Oxford and New York: Pergamon Press.

Pearce, Fred. 2013. "Flushed with Success: Human Manure's Fertile Future." *New Scientist*, February 16: 48–51.

Perkins, Dwight, ed. 1977. *Rural Small-scale Industry in the People's Republic of China*. (ATL 03-75) Berkeley: University of California Press.

Pollan, Michael. 2013. *Cooked: A Natural History of Transformation*. New York: The Penguin Press.

Pollard, Justin. 2010. *Boffinology: The Real Stories Behind Our Greatest Scientific Discoveries*. London: John Murray.

Pomerantz, Jay M. 2004. "Recycling Expensive Medication: Why Not?" *Medscape General Medicine* 6 (2): 4.

Porter, Roy. 2003. *Blood and Guts: A Short History of Medicine*. New York: W. W. Norton.

Raford, Noah, and Jason Bradford. 2009. "Reality Report: Interview with Noah Raford." Resilience.org, July 17. http://www.resilience.org/stories/2009-07-17/reality-report-interview-noah-raford.

Rawles, James Wesley. 2009. *How to Survive the End of the World as We Know It: Tactics, Techniques, and Technologies for Uncertain Times*. New York: Plume / Penguin.

Read, Leonard E. 1958. *I, Pencil: My Family Tree as Told to Leonard E. Read*. Irvington-on-Hudson, NY: The Foundation for Economic Education. Reprint, 1999.

Reilly, Desmond. 1951. "Salts, Acids & Alkalis in the 19th Century: A Comparison between Advances in France, England & Germany." *Isis* 42 (130): 287–96.

Rooney, Anne. 2009. *The Story of Medicine: From Early Healing to the Miracles of Modern Medicine*. London: Arcturus Publishing.

Rose, Alexander. 2010. "Manual for Civilization." *Blog of the Long Now*, April 6. http://blog.longnow.org/02010/04/06/manual-for-civilization/.

Rosen, Nick. 2007. *How to Live Off-Grid: Journeys Outside the System*. London: Bantam Books.

Ross, Bill. 2005. "Building a Radio in a P.O.W. Camp." BBC: WW2 People's War. http://www.bbc.co.uk/history/ww2peopleswar/stories/70/a4127870.shtml.

Rybczynski, Witold. 1980. *Paper Heroes: A Review of Appropriate Technology*. (ATL 01-11) New York: Anchor Press.

Sacco, Joe. 2000. *Safe Area Goražde: The War in Eastern Bosnia 1992–95*. Seattle, WA: Fantagraphics Books.

Schaefer, Bradley E. 2000. "The Heliacal Rise of Sirius and Ancient Egyptian Chronology." *Journal for the History of Astronomy* 31 (2): 149–55.

Schlesinger, Henry. 2010. *The Battery: How Portable Power Sparked a Technological Revolution*. Washington, DC: Smithsonian Books.

Schrock, Richard. 2006. "Nitrogen Fix." *MIT Technology Review*, May 1. http://www.technologyreview.com/notebook/405750/nitrogen-fix/.

Schwartz, Glenn M., and John J. Nichols, eds. 2010. *After Collapse: The Regeneration of Complex Societies*. Tucson, AZ: The University of Arizona Press.

Sella, Andrea. 2012. "Classic Kit—Kenneth Charles Devereux Hickman's Molecular Alembic." *Solarsaddle's Blog*, January 6. http://solarsaddle.wordpress.com/2012/01/06/classic-kit-kenneth-charles-devereux-hickmans-molecular-alembic/.

Seymour, John. 2009. *The New Complete Book of Self-sufficiency*. London: Dorling Kindersley.

Shapin, Steven. 1996. *The Scientific Revolution*. Chicago: The University of Chicago Press.

Shelley, Percy Bysshe. 1818. "Ozymandias." *The Examiner,* January 11.

Sherman, Irwin W. 2006. *The Power of Plagues*. Washington, DC: ASM Press.

Shirky, Clay. 2010. *Cognitive Surplus: Creativity and Generosity in a Connected Age*. New York: The Penguin Press.

Shuval, Hillel I., Charles G. Gunnerson, and DeAnne S. Julius. 1981. *Appropriate Technology for Water Supply and Sanitation: Night-soil Composting*. (ATL 17-389) Washington, DC: The World Bank.

Silverman, Steve. 2001. *Einstein's Refrigerator: And Other Stories from the Flip Side of History*. Kansas City, MO: Andrews McMeel Publishing.

Smith, Gerald. 2009. "The Chemistry of Historically Important Black Inks, Paints and Dyes." *Chemistry Eduction in New Zealand*, (May): 12–15.

Sobel, Dava. 1995. *Longitude: The True Story of a Lone Genius Who Solved the Greatest Scientific Problem of His Time*. New York: Walker & Company.

Solomon, Steven. 2011. *Water: The Epic Struggle for Wealth, Power, and Civilization*. New York: Harper Perennial.

Solomon, Susan, Gian-Kasper Plattner, Reto Knutti, and Pierre Friedlingstein. 2009. "Irreversible Climate Change Due to Carbon Dioxide Emissions." *Proceedings of the National Academy of Sciences* 106 (6): 1704–09.

Spinney, Laura. 1996. "Return to Paradise: If the People Flee, What Will Happen to the Seemingly Indestructible?" *New Scientist*, July 20: 26–31.

Standage, Tom. 2010. *An Edible History of Humanity*. New York: Walker & Company.

Stanford, Geoffrey. 1976. *Short Rotation Forestry: As a Solar Energy Transducer and Storage System*. (ATL 08-301) Conference on Energy and Agriculture, St. Louis, MO, June. Dallas: Greenhills Foundation.

Starkey, Paul. 1989. *Harnessing and Implements for Animal Traction: An Animal Traction Resource Book for Africa*. (ATL 06-294) Braunschweig and Wiesbaden: German Appropriate Technology Exchange (GATE) and Friedr. Vieweg & Sohn.

Stassen, Hubert E. 1995. *Small-Scale Biomass Gasifiers for Heat and Power: A Global Review*. World Bank Technical Paper Number 296. Washington, DC: World Bank.

Stein, Matthew R. 2008. *When Technology Fails: A Manual for Self-Reliance,*

Sustainability, and Surviving the Long Emergency. 2nd ed. White River Junction, VT: Chelsea Green Publishing.

Stern, Nicholas. 2006. *The Economics of Climate Change*. The Stern Review. HM Treasury.

Stern, Peter. 1979. *Small Scale Irrigation*. (ATL 05-217) London: Intermediate Technology Publications.

———, ed. 1983. *Field Engineering*. (ATL 02-71) Bourton on Dunsmore, UK: Practical Action Publishing.

Stoner, Carol Hupping, ed. 1973. *Stocking Up: How to Preserve the Foods You Grow, Naturally*. (ATL 07-292) Emmaus, PA: Rodale Press.

Strauss, Neil. 2009. *Emergency: This Book Will Save Your Life*. New York: It Books / HarperCollins.

Strawbridge, Dick, and James Strawbridge. 2010. *Practical Self Sufficiency: The Complete Guide to Sustainable Living*. London: Dorling Kindersley.

Sutton, Christine. 1986. "The Impossibility of Photography." *New Scientist*, 25 December 1986 / 1 January 1987: 40–43.

Tainter, Joseph A. 1988. *The Collapse of Complex Societies*. New Studies in Archaeology. Cambridge, UK: Cambridge University Press.

Thwaites, Thomas. 2011. *The Toaster Project: Or a Heroic Attempt to Build a Simple Electric Appliance from Scratch*. New York: Princeton Architectural Press,.

United Nations Development Fund for Women. 1988. *Cereal Processing*. (ATL 06-299) New York: UNIFEM.

United States Army. 2002. *Survival (Field Manual 3-05.70)*. Fort Belvoir, VA: Army Publishing Directorate.

United States Department of Agriculture, 1974. *Wood Handbook: Wood as an Engineering Material*. (ATL 25-662) Madison, WI: United States Forest Service Forest Products Laboratory, USDA.

———, 2008. *Cuba's Food & Agriculture Situation Report*. Office of Global Analysis, Foreign Agricultural Service, USDA.

United States Department of Energy. 1980. *Fuel from Farms: A Guide to Small-scale Ethanol Production*. (ATL 19-417) Golden, CO: Solar Energy Research Institute United States Department of Energy.

Usher, Abbott Payson. 1982. *A History of Mechanical Inventions*. Rev. ed. (1954) Mineola, NY: Dover Publications. First published 1929 by Harvard University Press.

van Lengen, Johan. 2008. *The Barefoot Architect: A Handbook for Green Building*. Bolinas, CA: Shelter Publications.

van Vuuren, D. P., M. Meinshausen, G-K. Plattner, et al. 2008. "Temperature Increase of 21st Century Mitigation Scenarios." *Proceedings of the National Academy of Sciences* 105 (40): 15258–62.

van Winden, John. 1990. *General Metal Work, Sheet Metal Work and Hand Pump Maintenance*. (ATL 04-134) TOOL Foundation.

Vigneault, François. 2007. "Papermaking 101." *Craft* (Winter): 132–37. http://makezine.com/projects/papermaking-101/.

Vogler, Jon. 1981. *Work from Waste: Recycling Wastes to Create Employment.* (ATL 33-804) London: Intermediate Technology Publications.

———. 1984. *Small-Scale Recycling of Plastics.* (ATL 33-799) London: Intermediate Technology Publications.

Volunteers in Technical Assistance. *Using Water Resources.* 1977. (ATL 12-327) Mount Rainier, MD: VITA.

Ware, Mike. 1997. "On Proto-photography and the Shroud of Turin." *History of Photography* 21 (4): 261–69.

———. 2002. "Luminescence and the Invention of Photography: 'A Vibration in the Phosphorus.'" *History of Photography* 26 (1): 4-15.

———. 2004. Mike Ware: "Alternative Photography." Website, http://www.mikeware.co.uk

Watson, Simon, and Murray Thomson. 2005. *Feasibility Study: Generating Electricity from Traditional Windmills.* Draft Final Report. Loughborough, UK: The Centre for Renewable Energy Systems Technology, Loughborough University.

Weisman, Alan. 2007. *The World Without Us.* New York: Thomas Dunne Books.

Wells, R. G. (interview by Brian James, Oral History Research Unit, Bournemouth University, UK, September 13, 1995). "Construction of Radio Equipment in a Japanese POW Camp." Oral History of Defense Electronics. Bournemouth, UK: Bournemouth University. http://histru.bournemouth.ac.uk/CHiDE/Oral_History_of_Defence_Electronics/r_g_wells.htm.

Werner, David, with Carol Thuman and Jane Maxwell. 2011. *Where There Is No Doctor: A Village Health Care Handbook.* Berkeley, CA: The Hesperian Foundation.

Weygers, Alexander G. 1973. *The Making of Tools.* (ATL 04-103) New York: Van Nostrand Reinhold Company.

———. 1974. *The Modern Blacksmith.* (ATL 04-108) New York: Van Nostrand Reinhold.

Whitby, Garry. 1983. *Glassware Manufacture for Developing Countries.* (ATL 33-792) Technical Papers 2. London: Intermediate Technology Development Group.

Wigginton, Eliot, ed. 1973. *Foxfire 2: Ghost Stories, Spring Wild Plant Foods, Spinning and Weaving, Midwifing, Burial Customs, Corn Shuckin's, Wagon Making and More Affairs of Plain Living.* (ATL 02-33) New York: Anchor Books.

Wingate, Michael. 1985. *Small-Scale Lime-burning: A Practical Introduction.* (ATL 25-675) London: Intermediate Technology Publications.

Winston, Robert. 2010. *Bad Ideas? An Arresting History of Our Inventions.* London: Bantam Press.

Wiseman, John "Lofty." 2010. *SAS Survival Handbook: The Ultimate Guide to Surviving Anywhere.* Rev. ed. New York: Collins.

Wood, T. S. 1981. *Simple Assessment Techniques for Soil and Water.* (ATL 05-213) New York: Coordination in Development, Environment and Development Program.

Wyndham, John. 1951. *The Day of the Triffids.* New York: Doubleday.

Yeo, Richard. 2001. *Encyclopaedic Visions: Scientific Dictionaries and Enlightenment Culture.* Cambridge, UK: Cambridge University Press.

Zalasiewicz, Jan. 2008. *The Earth After Us: What Legacy Will Humans Leave in the Rocks?* Oxford and New York: Oxford University Press.

ACKNOWLEDGMENTS

It goes without saying that while it is my name that appears on the front cover, this book would never have come into existence without the hard work and expertise of a great number of people helping me along the way. So starting from the beginning, with my awesome literary agent, Will Francis: Thank you, Will, for getting in touch back in 2008 after reading *Life in the Universe* and for all your guidance and encouragement over the years since, and, let's be honest, for outright hassling me into going beyond simply mulling over this concept in the back of my mind and actually researching and writing a book on it . . . Thanks too to Kirsty Gordon, Rebecca Folland, and Jessie Botterill in the Janklow & Nesbit agency offices in London for all their help, as well as PJ Mark and Michael Steger in New York.

Thank you to Stuart Williams at The Bodley Head and Colin Dickerman at Penguin Group (USA) for showing so much enthusiasm for the idea and for your faith in me actually pulling this ambitious project off. I'm enormously indebted to Colin, and especially Jörg Hensgen (The Bodley Head) for his unbelievably skillful and perceptive editing of my writing: any finesse in this finished book is from his exquisite craftsmanship, which has uncovered and polished a sculpture hidden within the roughly hewn block of stone I submitted as a first draft. Many thanks also to Akif Saifi and Mally Anderson for all her help and to Scott Moyers (Penguin), who seamlessly took over from Colin Dickerman. And a greatly appreciative bow to Katherine Ailes (The Bodley Head), particularly for all her efforts in securing such a stunning set of images to adorn these pages and bring the words alive. Thanks too to Maria Garbutt-Lucero and Will Smith (The Bodley Head) and Samantha Choy Park, Sarah Hutson, and Tracy Locke (Penguin) for your help with the publicity and marketing of the book.

The subject matter of this book is very eclectic, and has taken me far beyond the horizons of my own academic field of expertise. Conducting the research has brought me into contact with a hugely diverse range of people, and I have been constantly warmed by the extent to which people will go in offering their time and effort to help a stranger. These contributions have been utterly invaluable and include: replying to an out-of-the-blue e-mail with useful information and tip-offs of what else to look into; agreeing to be subjected to me picking their brains with a toddler-like series of whys, whats, and hows; helping with illustrations or reading through draft chapters to check for howlers; kindly and gently pointing out factual errors in the hardcover text; and generously spending hours sitting down with me and explaining slowly (and repeatedly!) the details and history of their own specialties. So a deep and heartfelt thank you to:

Paul Abel, Jon Agar, Richard Alston, Stephen Baxter, Alice Bell, John Bingham, John Blair, Keith Branigan, Alan Brown, Mike Bullivant, Donal Casey, Andrew Chapple, Jonathan Cowie, Thomas Crump, Sam Davey, John Davis, Oliver de Peyer, Klaus Dodds, Julian Evans, Ben Fields, Steve Finch, Craig Gershater, Vince Gingery, Vinay Gupta, Rick Hamilton, Vincent Hamlyn, Colin Harding, Andy Hart, Rebekah Higgitt, Tim Hunkin, Alex Karalis Isaac, Richard Jones, Jason Kim, Karol Kleszyk, James Kneale, Roger Kneebone, Monika Koperska, Nancy Korman, Paul Lambert, Simon Lang, Marco Langbroek, Pete Lawrence, Andrew Mason, Gordon Masterton, Rich Maynard, Steve Miller, Mark Miodownik, John Mitchell, Ginny Moore, Terry Moore, Francisco Morcillo, James Mursell, Jheni Osman, Matt Parker, Sam Pinney, David Pryor, Antony Quarrell, Noah Raford, Peter Ransom, Clive Reed, Carole Reeves, Alby Reid, Alexander Rose, Steven Rose, Andrew Russell, David Schwan, Tim Sammons, Andrea Sella, Anita Seyani, James Sherwin-Smith, Tony Sizer, William Slaton, Simon Smallwood, Colin Stuart, Frank Swain, Stefan Szczelkun, Ian Thornton, Thomas Thwaites, Phiroze Vasunia, Alex Wakeford, Mike Ware, Simon Watson, Andrew Wear, Kathy Whalen Moss, Sophie Willett, Emma Williams, Andrew Wilson, Peter Wilson, Lofty Wiseman, and Marek Ziebart.

If civilization ever does go belly-up, I would feel privileged to have any of you on my post-apocalyptic survival team!

Thank you to Max Richter, Arvo Pärt, Godspeed You! Black Emperor, M83, Tom Waits, Kate Rusby, and Jon Boden (your *Songs from the Floodplain* is quite possibly the best post-apocalyptic folk album in the genre . . .) for providing the soundtrack within my work bubble, and Nor and Fat Cat cafés for putting up with my long hours of mocha mainlining and lip-chewing while writing. Your pork belly sandwiches are the pinnacle of civilized society.

Thank you too to my family and friends who have smilingly endured my repetitive dinner table and pub chat on post-apocalyptic matters, or humored me on research adventures. The final and most important thanks are, of course, to my wonderful wife. Vicky has stoically supported me through this long process, quietly tolerating the many weekends lost to a grouchy husband hunched over the laptop and effortlessly picking up my mood after an evening home alone "doing background research" from bleak post-apocalyptic films and novels.

Page numbers in *italics* refer to illustrations.

electric vehicles, 206–8
electrochemical cell, 175
electrolysis, 212, 232–33, 234, 249
electrolyte, 174
electromagnetism, 174, 176–77, 184,
    220–25
electrons, 152–53, 174, 175, 183
elements, 233–34
    periodic table of, 234–35, 242
Eliot, T. S., vii, 10, 275
Encyclopédie (Diderot), 8, 19
energy and power, 15, 165–85
    batteries for, *see* batteries
    consumption per person, 166
    from fossil fuels, 31, 57–58, 105, 165,
        166, 190
    generation and distribution of, 178–85
    mechanical, 166–74
    solar, 13, 47
    steam engine for, 170n, 172–73, 182, 197,
        201, 277, 290–91
    steam turbine for, 182–83, 185, 206–7
    thermal, 104–9, 166, 172, 182
    water turbine for, 180–82, *181*
    waterwheel for, 166–68, *167*, 170, 171–72,
        178–79, 180, 203, 213, 276
    windmill for, 46, *169*, 170, 171–72, 178–80,
        179, 213
    see also electricity generators
Energy Return on Energy Invested
    (EROEI), 196n
engines:
    diesel, 188, 190
    gasoline, 187, 188
    internal combustion, 201–6, *204*, 208
    steam, 170n, 172–73, 182, 197, 201
Enlightenment, 276
enzymes, 81–82
equinoxes, 259, 261, 265
ethanol, 43, 89, 91, 120, 158, 159, 189–90, 202,
    206, 241
ether, 159, 163, 241
evolution, 279
explosives, 104, 110, 116, 118, 122, 235–38,
    242, 247

fabrics, *see* textiles
fats and oils, 112, 188
    animal, 191
    linseed, 219
    saponification of, 84, 112–13, 114–15, 211–12
    vegetable, 108–9, 188, 190

feather pen, 214
fermentation, 80, 84, 88, 90–91
ferrous sulfate, 241
fertilizers, 57, 58, 69, 70, 72, 73, 76, 77n, 110,
    121, 237, 238, 247–48, 250, 278–79
    manure, 61, 69, 70, 73–76, 237, 243,
        247, 250
Feynman, Richard, 9–10
fire:
    starting, 34–35
    using, 104–5, 123–24
fires, destructive, 28
firewood, 106, 165, 176
fishing, 198
flash paper, 238
flax, 96, 108, 211
Fleming, Alexander, 12, 162, 164
Florey, Howard, 163
flour, 39, 86, 87
flying shuttle, 101
food, 33, 38–41
    canned, 40, 92, 291
    cereal preparation, 86–91
    cooking, 79–81, 124
    growing, *see* agriculture
    poisoning from, 79–81, 84
    pottery vessels for, 80–81, 95,
        124–26
    spoilage of, 80, 161, 163
food preservation, 80, 81–85,
    91–95, 161
    by canning, 92, 291
    by drying, 82, 91–92
    by fermenting, 80, 84
    by pickling, 84, 89, 91–92, 118
    by refrigeration, 93–95
    by salting, 82–83, 91–92
    by smoking, 83, 91–92
fool's gold, 222
forceps, 149–50, *150*
fore-and-aft rigging, 198–99
forests, 27, 30, 106
fossil fuels, 31, 57–58, 105, 165, 166, 190
Foucault's pendulum, 256n
Four Quartets (Eliot), 275
foxglove, 155
Frank, Pat, 165
fuels, 41–42, 119, 124
    biofuels, 74–75, 119, 191, 206–7, 208
    gasification and, 191–93, *192*, *194*
    kerosene, 108–9
fungicide, 118

metalworking, 132–35, *133*, *134*, 195, 200
methane, 74, 191, 193, 249
methanol, 118–19, 190
metric system, 282–84
microbiology, 160–64
microorganisms, 80, 81, 279, 291
microphone, 225–26
microscope, 141, 143–44, 160–61, 288
milk, 39, 84, 85, 92
Millennium Seed Bank, 54
Miller, Walter M., Jr., 123
millet, *66*, 67, 86
mill pond, 168
millstone, 86, 95
mirror, 240*n*
Moldova, 3*n*
morphine, 156
Morse code, 221
mortars, lime, 126, 127–29
motor, car, 173

Napoleon I, Emperor, 282*n*
Napoleon III, Emperor, 233*n*
nature, 279, 287
    urban spaces reclaimed by, 26–30, *27*, 106
navigation, *see* location, determining
New Orleans, La., 20–21
New York City, 256*n*, 266–67, 268, 270
nitrates, 237, 238, 243, 247, 248
nitric acid, 121–22, 159, 238, 247, 250
nitrocellulose, 238
nitrogen, 69, 70, 72, 73, 75, 88, 115, 178, 238,
    247–51
    in Haber-Bosch process, 57, 232, 248–51
nitroglycerin, 238
nitrous oxide, 159, 250
Nobel, Alfred, 238
north pole, 263–65, *264*
North Star (Polaris), 263, 269
nuclear war, 22

oats, 60, *66*, 67
Off-Gridders, 48
oil, crude, 116, 119–20, 189, 279
oil lamps, 108–9, 184
oil paints, 219
oils, *see* fats and oils
Oklahoma City bombing, 250*n*
opium, 156
oral rehydration therapy (ORT), 148
Oryx and Crake (Atwood), 103
oscillator, 257–58, 271

oscillator circuit, 224–25
oxen, 195, 196
oxidizing agents (oxidants), 104, 236, 238
oxygen, 232, 233, 279
    explosives and, 236
"Ozymandias" (Shelley), 209

pain relief, 155–57, 158–59
paints, 118
    oil, 219
pandemic, 23
Pantheon, 29
paper, 210–13, *215*, 218, 232, 238, 243, 276
pasteurization, 92
pathology, 154
pellagra, 84*n*
Pelton turbine, 181, *181*, 182
pen, 214
pendulum, 283
    clock, 257–59, 271, 289–90
    Foucault's, 256*n*
penicillin, 12, 162, 163–64
periodic table, 234–35, 242
pesticides and herbicides, 57, 77*n*, 279
Petri dishes, 161, 162
pharmaceutical compounds, *see* medicines
phones, 3–4, 13, 50–51
phosphorus, 57, 75–76, 110
photography, 104, 114, 122, 235, 238, 239–42
    observing stars and, 263, 265
pickling, 84, 89, 91–92, 118
piezoelectric crystal, 226
piston, 173
pitch, 119
plastic bottles, for solar water disinfection,
    37–38
plastics, 49–50, 116, 118, 235*n*, 279
plate tectonics, 279
plow, 58, 60–61, *62*, 63, 195
Polaris, 263, 269
polyethylene terephthalate (PET), 50
poppy, 156
population, 24–25
potash, 17, 76, 114, 115, 120, 139, 212, 233, 234,
    237, 243–44, 249
potassium, 57, 75, 76, 114, 233, 234
    bitartrate, 226
    carbonate, 114
    hydroxide, 115, 190, 249
    nitrate (saltpeter), 237–38, 248, 250
    sodium tartrate, 226
potatoes, 60, 65, 67

transmission, car, 204
transport, 185, 187–208
    electric vehicles for, 206–7
    fuels for, 188–93
    gasification and, 191–93, *192, 194*
    keeping vehicles running, 188–95
    and loss of mechanization, 195–201, *197*
    powered, reinventing, 201–8
    roads for, 187–88
    rubber for, 193–95
trees, 17
triode, 229
trip hammer, 171, *171,* 179, 203, 276
tuberculosis, 147
turpentine, 119, 219–20
type setting, 216–18, *217*

urea, 115
urine, 73, 74, 115

vacuum, 284
vacuum tubes, 141–42, 153, 227–30
Varro, Marcus Terentius, 161*n*
vinegar, 84, 118, 120
violence and crime, 20–22
vitamin B3, 84*n*
vitamin D, 72, 84–85
vitriol, 120
voltaic pile, 175, 178

walkie-talkies, 51
Washington, DC, 45
waste, human, 73–75
Waste Land, The (Eliot), vii, 10
water, 81, 103
    for apartment buildings, 44–45
    destruction caused by, 28–30
water, drinking, 33, 36–38, 109–10, 124, 148
    oral rehydration therapy, 148
water clock, 205, 206, 254

water turbine, 180–82, *181*
    steam, 182–83, 185
waterwheel, 166–68, *167,* 170, 171–72, 178–79,
    180, 203, 213, 276
Watt, James, 170*n*
weaving, 98, 277
weaving loom, 98–100, *99,* 101–2
welding, 131–32
wet collodion process, 241
wheat, 53–54, *55,* 60, *66,* 67, 68, 70, 71,
    73, 86, 89
wheelbarrow, 12, 276
Wikipedia, 9
willow, 155, 157
wind, 168–70
windmill, 46, *169,* 170, 171–72, 178–80,
    179, 213
windows, 140, 141
winnowing, 68
Wiseman, John "Lofty," 33
wood, 17, 123–24
    ashes, 17, 76, 113–14, 115
    charcoal from, 106–7, 116, 124, 135, 184
    firewood, 106, 165, 176
    gasification of, 191–93
    paper from, 211–13
    pyrolysis of, 116–20, *117,* 192
    smoking food with, 83, 91–92
wood alcohol (methanol), 118–19, 190
wool, 96–98
World Health Organization, 37
World War II, POW radios in, 226–27
writing, 210–14, 215
Wyndham, John, 53
X-rays, 12, 141, 153, 221, 239

yeast, 89–91
yogurt, 84, 85, 88

Zeer pot, 93

CREDITS

Page 61: Simple farming tools, illustration by Bill Donohoe.

Page 62: Complex farming tools: plough from *Lexikon der gesamten Technik* by Otto Lueger; harrow, seed drill, and plough action from *Meyers Konversationslexikon* (1905–1909) by Joseph Meyer. All reproduced courtesy of www.zeno.org.

Page 66: Cereal crops from *Meyers Konversationslexikon* (1905–1909) by Joseph Meyer, reproduced courtesy of www.zeno.org; page design by Bill Donohoe.

Page 69: Mechanical reaper from *Meyers Konversationslexikon* (1905–1909) by Joseph Meyer, reproduced courtesy of www.zeno.org.

Page 97: Spinning wheel from *The Wonderful Story of Britain: The New Spinning Machine* by Peter Jackson (1922–2003) / Private Collection / © Look and Learn / The Bridgeman Art Library.

Page 99: Loom © Science Museum / Science & Society Picture Library. All rights reserved.

Page 117: The pyrolysis of wood: (TOP) drawing of retort for wood distillation taken from p.12 of *Wood products: distillates and extracts* by Dumesny and Noyer (1908); (BOTTOM) diagram by author.

Page 133: Rudimentary foundry, photographs reproduced by kind permission of David J. Gingery Publishing, LLC.

Page 134: Lathe © Science Museum / Science & Society Picture Library. All rights reserved.

Page 137: Blast furnace, illustration by Bill Donohoe.

Page 150: Birthing forceps, reproduced courtesy of Historical Collections & Services, Claude Moore Health Sciences Library, University of Virginia.

Page 167: Overshot waterwheel, fig. 56 on p.109 from *Flour for Man's Bread: A History of Milling* by John Storck and Walter Dorwin Teague (UMP, 1952), copyright 1952 by the University of Minnesota, renewed 1980.

Page 169: Self-orienting turret windmill from "Wind powered factories: history (and future) of industrial windmills," in Low-Tech Magazine, article © Kris De Decker (edited by Vincent Grosjean).

Page 171: Crank and cam, illustration by Bill Donohoe.

Page 179: Charles Brush's electricity-generating windmill, reproduced courtesy of the Western Reserve Historical Society.

Page 181: Pelton turbine, reproduced courtesy of the A. P. Godber Collection, Alexander Turnbull Library.

Page 192: Gas-bag bus © Scotsman Publications.

Page 194: Wood gasifier car © Mike LaRosa.

Page 197: Makeshift horse-drawn trap © Sean Caffrey / Getty Images.

Page 204: Internal combustion engine, illustration by Bill Donohoe.

Page 217: Mold for type casting, author's photograph.

Page 228: Simple radio receiver: wiring diagram (TOP) by Bill Donohoe; photograph (BOTTOM) reproduced courtesy of Tim Sammons.

Page 246: Soda plant: photo (TOP) Solvay Process Company's works, Syracuse, N.Y. from the Library of Congress; diagram (BOTTOM) by author.

Page 258: Mechanical clock from *Meyers Konversationslexikon* (1905–1909) by Joseph Meyer, reproduced courtesy of www.zeno.org.

Page 263: Barnard's star, diagram by author.

Page 264: Shifting of the celestial North and South poles, diagram by author.

Page 269: Sextant, from p.1932 of *Webster's New International Dictionary of the English Language* (1911 edn).